Art of War Gift Books

Mastering Strategy Series

^{Sun Tzu's} The Art of War Plus The Warrior's Apprentice
A first book on strategy for the novice.

^{Sun Tzu's} The Art of War Plus The Ancient Chinese Revealed
See the original! Each original Chinese character individually translated.

^{Sun Tzu's} The Art of War Plus Its Amazing Secrets
Learn the hidden secrets! The deeper meaning of Sun Tzu explained.

The Warrior Class: 306 Lessons in Strategy
The complete study guide! Simple lessons in strategy you can read any time.

Career and Business Series

^{Sun Tzu's} The Art of War Plus The Art of Career Building
For everyone! Use Sun Tzu's lessons to advance your career.

^{Sun Tzu's} The Art of War Plus The Art of Sales
For salespeople! Use Sun Tzu's lessons to win sales and keep customers.

^{Sun Tzu's} The Art of War Plus The Art of Management
For managers! Use Sun Tzu's lessons on managing teams more effectively.

^{Sun Tzu's} The Art of War Plus Strategy for Sales Managers
For sales managers! Use Sun Tzu's lessons to direct salespeople more effectively.

^{Sun Tzu's} The Art of War Plus The Art of Small Business
For business owners! Use Sun Tzu's lessons in building your own business.

^{Sun Tzu's} The Art of War Plus The Art of Marketing
For marketing professionals! Use Sun Tzu's lessons to win marketing warfare.

Life Strategies Series

^{Sun Tzu's} The Art of War Plus The Art of Love
For lifelong love! *Bing-fa* applied to finding, winning, and keeping love alive.

^{Sun Tzu's} The Art of War Plus The Art of Parenting Teens
For every parent! Strategy applied to protecting, guiding, and motivating teens.

Current Events Series

^{Sun Tzu's} The Art of War Plus Strategy against Terror
An examination of the War on Terror using Sun Tzu's timeless principles.

Audio and Video

Amazing Secrets of *The Art of War*: Audio
1 1/2 Hours 2-CD set

Amazing Secrets of *The Art of War*: Video
1 1/2 Hours VHS

孫子兵法

Sun Tzu's

THE
ART
OF
WAR

for

Sales Force Success
Strategy for Sales Managers

孫子兵法

Sun Tzu's

THE
ART
OF
WAR

for

Sales Force Success
Strategy for Sales Managers

by Gary Gagliardi

Science of Strategy
Clearbridge Publishing

Published by
Clearbridge Publishing

FIRST EDITION
Copyright 1999, 2001, 2005 Gary Gagliardi

Manufactured in the United States of America.
Interior and cover graphic design by Dana and Jeff Wincapaw.
Original Chinese calligraphy by Tsai Yung, Green Dragon Arts, www.greendragonarts.com.

Publisher's Cataloging-in-Publication Data
Sun-tzu, 6th cent. B.C.
    [Sun-tzu ping fa, English]
    Strategy for sales managers / Sun Tzu and Gary Gagliardi
    p. 194 cm. 23
    Includes introduction on strategy of Sun Tzu and topical index
    ISBN 978-1-929194-55-1
    1. Selling. 2. Sales management. 3. Sales promotion. 4. Military art and science - Early works
to 1800. I. Gagliardi, Gary 1951— . II. Strategy for Sales Managers
HF5438.5.S86 2005
658.8 /1 21 —dc21

        Library of Congress Catalog Card Number: 2005900950

Clearbridge Publishing's books may be purchased for business, for any promotional use,
or for special sales. Please contact:

Clearbridge Publishing
2829 Linkview Drive, Las Vegas, NV 89134
www.scienceofstrategy.com
gagliardi.gary@gmail.com

Contents

The Art of War Plus
Strategy for Sales Managers

Foreword

An Unstoppable Sales Force

When I first discovered Sun Tzu's *The Art of War*, I was an average salesperson or maybe a tiny bit worse than average because I am naturally a little lazy. Despite my limitations, after I began mastering Sun Tzu's strategic principles, I was consistently the best salesperson in every company I joined. The result was that I was promoted into sales management. Though good salespeople don't necessarily make good sales managers, I was fortunate. I found that the strategic understanding that I had gained from studying Sun Tzu was even more valuable in a management role. I was working for Bic Pen at the time, and in my first year I won the company's zone manager of the year award, which included a nice grandfather clock that I still own.

Within a few years, I had my own company and my own sales force. The first thing I did was write an adaptation of Sun Tzu's principles for my salespeople, a book I called *The Art of Sales*. Again, the effect was dramatic. Our software company began to grow at a rate of 40 percent per year, twice appearing on the *Inc.* 500 list of fastest-growing companies in America. In addition, we began winning a number of business awards such as the Blue Chip Quality Award from the U.S. Chamber of Commerce.

As I began to hire sales managers away from our larger competitors, I discovered that though they had certain skills and industry knowledge, they lacked the strategic training that would have allowed them to succeed consistently. Unfortunately, I also discovered that, too often, their past success made it difficult for them to change their habits. In the end, I used these older experienced sales managers for their industry contacts, but I trained younger, hungrier managers to replace them—ones who knew how to think more strategically.

As our company became more successful, magazines such as *PC Week* began writing about our use of Sun Tzu's principles in business. Our customers, companies such as AT&T, GE, and Motorola, began asking me to speak on classical strategy, training both their salespeople and sales managers. When I sold my software company in 1997 and became financially independent, I continued to train sales departments to think more strategically. Companies such as IBM, 3M, and American Express have purchased tens of thousands of copies of our sales adaptation of Sun Tzu's *The Art of War* over the years.

A similar book for sales managers is a natural extension of this work. Sales managers are the perfect audience for Sun Tzu's lessons. Salespeople are officers on the front lines of battle, but sales managers are the business world's generals. All officers need to understand strategy, but Sun Tzu wrote *The Art of War* for the commanding general.

Strategies for Sales Managers tackles the broader, longer-term strategic issues that sales managers must address because salespeople cannot. Salespeople manage customer relationships, but sales managers must manage salespeople, a much more daunting task. The principles in this book for sales managers can be used in conjunction with *The Art of Sales*, giving your whole sales department a consistent understanding of sales strategy.

What is the science of strategy for a sales manager? Though we commonly use the term "strategy" as a loose synonym for planning or even a good idea, Sun Tzu teaches that strategy is a process, not just a plan. The process is scientific because it produces consistent results. Sun Tzu taught that planning alone does not work because competitive environments are too complex, fast-changing, and unpredictable for rigid, long-term plans. Instead, he taught a simple process for creating reproducible results under a wide variety of different and uncontrollable conditions.

Sales managers must make the right decisions at the right time. Sun Tzu's *The Art of War* offers a clear framework for making those competitive decisions. Instead of focusing on defeating opponents, strategy teaches that success is a matter of developing strategic positions that can be easily defended and advanced over time. Sun Tzu taught that our opponents leave us openings, but that we have to learn how to recognize these opportunities. We call Sun Tzu's strategy "winning without conflict" because it espouses leveraging your company's unique position against your opponents' weaknesses.

Sales management is the most difficult form of management. Salespeople are strong-willed and independent. Managing them is as difficult as herding cats. Selling is a highly emotional, challenging, and well-compensated job. It attracts people who are aggressive and self-reliant. Your job as a sales manager is to keep these people excited, motivated, and still under some form of control. Fortunately, Sun Tzu's lessons on managing people were designed to get the best out of officers and soldiers, another group of aggressive people in a highly emotional and difficult job.

To be a sales manager, you must be a superb salesperson yourself. An ordinary manager can excel at management alone, without excelling at the various tasks under his or her supervision. As a sales manager, you are the most important salesperson in your organization. Even if you don't get directly involved with every sale, you must understand the sales process, sales psychology, and sales techniques better than anyone else in your organization.

Every salesperson must think strategically, but as a sales manager you must be a true strategist. Strategy is decision-making tools. *The Art of War* offers a distinct, nonintuitive system for decision-making. It solidifies the vague idea of a strategy into a clear, well-defined set of principles. The book teaches that only a few key factors influence the outcome of your efforts. Success goes not to the strongest or most aggressive but to those who best understand their situation and what their alternatives really are. When you have mastered Sun Tzu's strategic principles, you will be able to almost instantly analyze competitive situations, spot opportunities, and make the appropriate decisions.

As a sales manager, you must see into the future. You must not only know how many sales will close by the end of the month, you must able to analyze your sales pipeline to predict what your sales volume will be in the future. Sun Tzu's principles give you very specific tools for predicting which sales you will win and which you will lose. You must teach your salespeople that they don't have time to pursue every possible opportunity. At its heart, *The Art of War*'s strategic system is about calculating the odds of winning and allotting your limited resources to maximize results.

As a sales manager, you play a key role in communication within the company. You must get feedback from your salespeople about how well the company's products and services are being accepted in the marketplace. You get solid information from the front lines to pass on to the people managing product development, marketing, and inventory planning. In our adaptation for sales managers, the basics of strategy are tailored to help you in your role as a communicator, persuader, and motivator.

As in our other books in the *Art of War Plus* Book Series, we present our adaptation for sales managers side by side with our complete translation of the original text of *The Art of War*. We suggest that in reading this work, you read both texts and not just our version for sales managers. Though phrased in military terms, Sun Tzu's original text lays out the elements of strategy in a very sophisticated manner. (It is so sophisticated that we have written a whole series of books explaining it in more detail.) Our sales management version, in contrast, is designed to make it easier to use those principles for the specific task of managing salespeople.

Each chapter in this book is preceded by a one-page introduction that explains how the strategic concepts covered in the chapter apply to sales management. Sections within each chapter are numbered to indicate where Sun Tzu introduces a new principle. These principles are simple and direct, and we adapt them line by line from their original military terminology to the world of sales management. The underlying concepts that these principles are built on, however, are rich and complex. The relationships between these basic principles are explained in some detail in the Introduction that follows.

Sun Tzu wrote precisely and succinctly, offering his ideas in a very compact format. Think of Euclid's *Geometry*. Both Euclid and Sun Tzu offer a set of basic concepts that build upon one another. Sun Tzu's principles, like Euclid's, have a wide variety of specific applications. Sales managers and salespeople can apply them to their very different tasks.

Strategies for Sales Managers offers a unique viewpoint. It proposes that running a successful sales team depends on your ability to motivate your salespeople to think strategically themselves. Strategy demands a focus on establishing and advancing winning positions rather than simply attacking the competition. It is a philosophy of continuous improvement. This focus on positioning and thinking about your position is what makes Sun Tzu's work so valuable.

In Sun Tzu's view, the secret to success is not just winning battles. It is developing positions that enable you to win without fighting battles. Battles are expensive. If you can guide your sales-people to win sales quickly and economically, your whole organization will benefit. As your sales team advances your position by winning customers, Sun Tzu teaches you how to hold onto the territory that you have won. The key to developing your marketplace is using each sale to win future sales more easily. As a sales manager, you must be wary of "victories" that consume your time and energy but that fail to position your company for long-term success in building a dominating market position.

Sales managers must define the standard sales process for their salespeople. However, Sun Tzu teaches that having a good sales process is not good enough. You must continually improve that sales process. This means giving your salespeople a certain amount of freedom so that they can experiment in the sales process. When one salesperson discovers a new technique that works, it is your job to spread that technique throughout your sales organization. This process of improving your process never ends. Sales markets are dynamic. Your sales organization must change faster than the market in order to advance its market position.

As a sales manager, you must work with your organization's marketing team to identify the best possible customers for your product or service. This is what Sun Tzu calls picking the right battlegrounds. The task of identifying the most profitable groups of customers is difficult. Salespeople often make the mistake of thinking that they just need more prospects. As a sales manager, you must teach them that sales volume depends on a complex combination of factors: the number of prospects, the time it takes to close sales, the size of the potential sales, the potential for reorders, and so on. Sales is a numbers game that requires both efficiency and effectiveness. As a sales manager, your job is to direct your salespeople so that they are always working at their peak potential.

Strategy is a method of adapting to change. You must be opportunistic as a sales manager. You don't create opportunities. You can defend your existing position from attack, but you must continuously examine the marketplace for new openings for your products. The challenge is recognizing these opportunities when they present themselves, and, once you recognize them, you must have the confidence to act. Sales management often requires watchful patience. At other times, sales management requires instant action. Sun Tzu argues that opportunities are always abundant (every problem creates an opportunity) but that opportunities can be difficult to recognize and act upon.

Sun Tzu's view of competition is knowledge-intensive. He sees success going to the person who is the most knowledgeable. Sun Tzu's focus on information is so clear that he devotes his final chapter, USING SPIES, to it. In the sales management version, this chapter is adapted as THE POWER OF INFORMATION. Strategically, there is no substitute for good information. Knowledge means having better information than anyone else. For a sales manager, it means knowing what your salespeople are doing, what your customers are doing, and what your competitors are doing every day of the week.

The utility of Sun Tzu's principles means that you apply them in different ways in different situations. For this reason, you should read and reread this book at least once a year. The lessons you take away in one situation will be different than those you take away in a different situation. With every reading, you will develop more insight into Sun Tzu's methods and your own situation. As your situation changes, different parts of the book will become more important. This work is organized so that the broadest and longest-term issues, such as strategic planning, are addressed in the initial chapters. Later chapters tend to focus on the special challenges encountered under specific conditions.

Despite its relatively short length, this book contains more valuable information about good sales management practices than other books two or three times its size. Do not expect to appreciate all of its principles in one reading. Time spent studying Sun Tzu's system is always time well invested.

Reading and rereading this book is simply the first step in mastering the warrior's world of strategy. Sun Tzu's system is sophisticated and deep. Much of its sophistication is not readily apparent simply from reading the text. For in-depth study of Sun Tzu's methods, we have created a whole series of books on mastering strategy. We strongly recommend that you (and perhaps your best salespeople) take the time to read *The Warrior Class*, which will further develop your strategic knowledge and skills.

I started my career as a salesperson and as a sales manager, and I have never stopped being a salesperson and a sales manager. Success in selling is addictive. There is an excitement that comes from getting orders that few other experiences in life can equal. After I sold my software company and retired, I missed that excitement. This book and all the many books in our *The Art of War Plus* Book Series are the result of my love affair with the thrill of creating sales.

Gary Gagliardi, 2005

† † †

孙子兵法

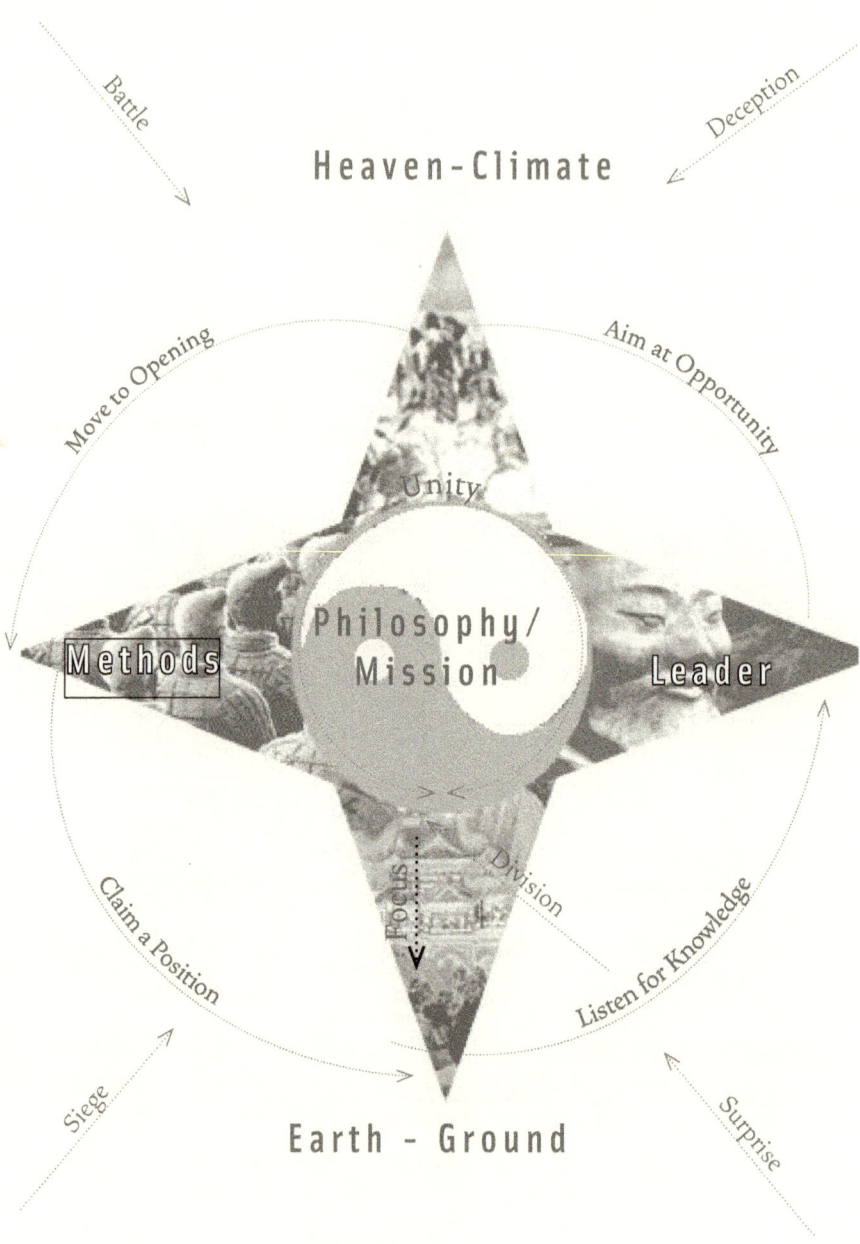

Heaven-Climate

Battle

Deception

Move to Opening

Aim at Opportunity

Unity

Methods

Philosophy/
Mission

Leader

Focus

Division

Claim a Position

Listen for Knowledge

Siege

Surprise

Earth - Ground

Introduction

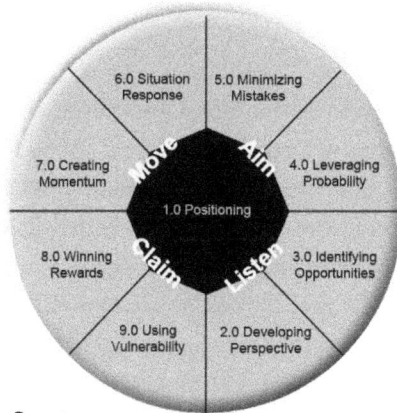

Sun Tzu's Strategic System

This book is for sales managers who want to introduce themselves
to the basic principles of Sun Tzu's competitive strategy and learn
to apply them to making better competitive decisions. If you are
new to Sun Tzu's strategic principles, you will find *The Art of War*
and The Art of Management easier to understand if you first familiar-
ize yourself with a few basic concepts, metaphors, and analogies.
Sun Tzu's system is based on the traditions of Chinese science and
philosophy and written from that perspective. Those traditions
organized ideas around five elements (see facing page) that date
back to the *I-Ching* and nine skills (see illustration above), derived
from the Chinese *Bagua,* that symbolized eight directions of movement plus its center. This introduction gives you an overview of
these key components of this strategic system.

As defined by Sun Tzu, strategy is not a system of planning.
Planning, in the sense of prioritizing a list of activities, works in
controlled environments where you can know how others will
respond to your decisions. Competitive strategy works in environments where your decisions collide with the decisions of others,
creating conditions that no one planned.

Most sales managers are confused about the competition. This
confusion often arises from two false dichotomies. The first is
between competition and cooperation, thinking that since cooperation means working together, competition means working

against others. The second false dichotomy is between competition and production, thinking that because production is productive, competition must be destructive. What people fail to see is that competition is essential to both cooperation and production.

Competition exists everywhere there are comparisons among alternative choices. All alternatives choices are "at war" with each other in the sense Sun Tzu used the idea. To put it more nicely, alternatives are in competition with each other. Cooperation requires competition because we must choose our partners. So potential partners are in competition with each other. Production requires competition because we must choose what to produce, which designs to use, what to sell, how to sell it, how to construct the supply chain, and so on. All these alternatives are in competition with each other.

The point of Sun Tzu's work is that, although competition, cooperation, and production all work together to produce value, they demand very different skill sets. Good competitive decisions are not made in the same way that good planning decisions are made. While planning depends on prioritization and organization, competitive decisions depend on agility and positioning, moving quickly and decisively to positions of strength.

Sun Tzu taught that in making these competitive choices, success is not a matter of winning fights with other people. Instead success depends on building and advancing strategic positions. A "position" is defined by Sun Tzu as what people actually compare in making competitive choices. The basis of conflict in competition is trying to weaken opposing positions. Sun Tzu's taught that conflict is inherently destructive, weaken the positions of everyone involved.

The focus of Sun Tzu's system is to create positions that others cannot attack and that ideally they want to join. Sun Tzu teaches that a general who fights a hundred battles and wins a hundred battles is not a good general. A good general is one who finds a way to win without fighting a single battle. Strategy teaches that you win by building the right positions and advancing those positions while avoiding conflict.

As a sales manager, your job is decision-making and problem-solving. According to Sun Tzu, all problems in competition are opportunities in disguise. If there were no problems, there would be no opportunities for improvement. More to the point, ff there were no problems, organizations would not need managers to make decisions on a day-to-day basis.

However, you cannot be successful in competition simply by tackling every problem. You must make decisions that advance your sales proposition. As you advance your sales proposition—and the position of your organization—you solve most problems by leaving them behind. The solution to most problems is identifying the weakness in your position that creates the problem and moving, to a sales position without that weakness.

This perspective clarifies many competitive situations in a way that nothing else does. Sun Tzu's strategy insists that you make the most of your strengths to compensate for your weaknesses. Your strengths and weaknesses do not come from your situation alone, but from your relative position within the marketplace. To make the right decisions, you must understand the key elements that define the strategic positions of your company and your competitors.

Sun Tzu uses five elements to define all positions within a competitive environment. Understanding positions is the first skill that his strategic system teaches. These five elements—philosophy/mission, heaven/climate, earth/ground, the leader, and methods—provide the backbone of a strategic approach to selling. All the other parts in his system—the nine skills, the four steps to advancing a position, the five faults of a leader, the six weaknesses of organizations, the six types of opportunities, the nine types of situations that arise in a campaign, and so on—flow from a good analysis of your relative position based on these five factors or elements.

The first element defining a position is called *philosophy* in the original text but it is better described in English as "the mission." A mission is the core of every position. It consists of the shared goals around which a team is built and the role your group exists to fulfill in the larger environment. Every department within an organization has its own mission, and the organization as a whole serves a larger mission. A clear idea of your mission within this larger purpose provides your group with its unity and focus.

Mission is the basis of two internal characteristics the Sun Tzu describes: *uniting* and *focusing* as creating competitive strength. Though these concepts are separate in English, in Chinese they are closely connected. In Sun Tzu's system, both arise directly from your shared goals or mission. *Uniting* holds the organization together. *Focusing* concentrates efforts in a single external goal. Sales teams lacking unity and focus from the element of mission are inherently weak..

The next two key elements define position within the larger environment. The importance of the environment is the great insight of Sun Tzu's work. He divides the environment into two opposite and yet complementary components, *heaven (climate)* and *ground (earth)*. Heaven and ground define the time and place of your position.

Heaven is the meaning of the Chinese character Sun Tzu used to describes time in terms of uncontrollable and unpredictable change. This climate arises from trends that change over time. Sales cycles are the most obvious trends in the sales environment, but every industry and organization has its own sales cycle and market climate. In today's business world, the pace of change erodes all existing sales positions and, at the same time, creates opportunities to advance your position. As a sales manager, you must focus your sales force on the right changes to improve your sales position.

Ground is the economic foundation on which your strategic position is based. It is both where you compete and what you compete for. It other words, it is the support of customer through sales. Unlike the market climate, which changes in ways largely beyond our control, your ground, your c\customer base, is determined by your own decisions. Choosing the prospects you pursue, how you position yourselves against the competition to win them, and utilizing them to generate future sales are part of choosing the right sales ground.

The *leader* is the next key factor in a strategic position. Decision-making is the unique responsibility of a leader in Sun Tzu's system. All sales managers are, by definition, leaders because they make decisions. Leadership is the realm of individual action and character. A sales manager needs to make the right decisions quickly. You need to choose the right sales team and decide who is doing the job and who is not.

Methods are the systems of the sales organization. Methods are, by definition, the realm of group action. A sales manager must utilize and build on existing systems to makes the organization's sales actions more effective. A sales manager needs a deep understanding of all the systems in an organization that affect his or her mission.

Much of Sun Tzu's work is about "attack," by which he means not fighting others but moving into new areas to advance your position. Advancing your position is a matter of a *leader* making good decisions to develop your team's *methods* to take advantages of *climate* changes to secure better *ground.* to meet your *mission.* Though it is the basis for the entire work, Sun Tzu describes the basics of these five factors in the first few pages of his work.

Once you see your strategic position clearly in terms of these five factors, you advance your position. The balance of the book addresses specific methods for advancing a position. However, all this information can be boiled down to four simple steps. These steps are *listening for knowledge, aiming at opportunities, moving to openings*, and *claiming a position*. Every advance requires all four steps. If you miss a step, the process is more likely to create problems than to solve them.

Knowledge comes from understanding your sales ground, which requires *listening. Aim* means seeing how changing trends create an *opportunity* to advance. *Moving* means the ability to change methods to take advantage of an *opening. Claiming* means reaping the rewards from a new ground *position.*

Each of these four steps leads, naturally to the next in an endless cycle of sales improvement. The more you learn about your ground, the more you need to identify new opportunities. Aiming at a new opportunity necessitates moving to new sales methods. Moving must give you a new position that your sales people can claim. Claiming new ground creates new opportunities to listen and learn. Even if your attempted advance fails to yield profitable new ground, it cannot fail to generate new knowledge, which is the basis of the next cycle and your inevitable success.

Consciously or unconsciously, you go through this cycle every time you advance your sales position. When a decision is unsuccessful, it is simply because one of these four steps was not properly executed.

These four steps can be further broken down into a list of the eight strategic skills. Remember, Sun Tzu's first skill is *understanding positions*. Listening requires Sun Tzu's second key strategic skill of an outside *developing perspective* on your position and his third skill of using the change of climate for *identifying opportunities*. Aiming requires Sun Tzu's fourth and fifth skills, *leveraging probability* and *minimizing mistakes*. Moving requires his six and seventh skills of *situation response* and *creating momentum*. Claiming requires his eighth and ninth skills of *winning rewards* and *understanding vulnerabilities*. All of these skills are covered in detail in our *Sun Tzu's Playbook*, which is divided into nine volumes, one for each skill. Articles from these nine volumes are referenced at the end of each chapter to connect the concepts presented in that chapter to the larger system.

The Art of War is a complete guide to executing these four steps in a wide variety of situations. However, much of it is written in a kind of code. These four steps are usually referenced in terms of metaphors. Listening for knowledge is referenced as sound. Thunder, music, and drums are all metaphors for listening. Aim is described as vision. Colors, lightning, and so on are all metaphors for foresight. Moving is marching. Claiming a position is variously described as gathering food, building, eating, digging in, and so on. We make all of these ideas easier to understand by adapting these metaphors into more easily understood management terms.

Just as these four steps are defined in terms of the five factors of a position, other strategic responses are defined in terms of these four steps. *Surprise* undermines knowledge. *Deception* confuses aim. *Battle*—which means meeting a challenge, not necessarily conflict— counters movement. *Siege* tries to overturn a position.

For a picture of Sun Tzu's system of five elements, four steps, and four responses, you can refer to the diagram that precedes this introduction.

If you find this introduction interesting, when you finish this book you may want to buy an overview of Sun Tzu's system written from a business perspective, *9 Formulas for Business Success: The Science of Strategy*. If you are looking for a more detailed analysis, you should explore the different volumes of *Sun Tzu's Art of War Playbook*, that provides a similar but much more extensive step-by-step guide to using Sun Tzu's system in business.

✦ ✦ ✦

Chapter 1

計

Analysis: Strategic Sales Position

Your sales organization has a unique sales position. You already have a lot of information about the position, but how well is that information organized? Are you overlooking some important aspects of your sales position? The framework provided in this first chapter gives you the tools not only for organizing what you already know but for expanding on that knowledge.

The first section of this chapter explains the five basic factors that define your strategic position. These five factors are simple categories that allow you to organize your knowledge about your situation.

Strategy teaches that you must always question your position. No position is good or bad in itself. The strengths and weaknesses of your position are defined by how you stack up against your competition. You cannot take your position for granted in a competitive situation in which positions constantly change. You can only discover opportunities if you are open to new points of view, especially those from outside your organization.

Strategy teaches that there is always a difference between perception and reality. Successful sales managers are those who can control perceptions while seeing through the deceptions of their competitors.

Your job as a sales manager is to provide an objective viewpoint for your salespeople. Your analysis must keep salespeople out of losing sales situations and guide them to winning ones.

Analysis

SUN TZU SAID:

This is war. 1
It is the most important skill in the nation.
It is the basis of life and death.
It is the philosophy of survival or destruction.
You must know it well.

[6]Your skill comes from five factors.
Study these factors when you plan war.
You must insist on knowing your situation.

1.	Discuss philosophy.
2.	Discuss the climate.
3.	Discuss the ground.
4.	Discuss leadership.
5.	Discuss military methods.

[14]It starts with your military philosophy.
Command your people in a way that gives them a higher
shared purpose.
You can lead them to death.
You can lead them to life.
They must never fear danger or dishonesty.

Strategic Sales Position

1 Sales managers must be warriors.
As a sales manager, you have a pivotal role in your organization.
You determine your organization's success or failure.
Your understanding is the path to growth or decline.
You must analyze your position.

Five key elements define your sales position.
Consider these factors when you plan to advance your position.
To know your company's strategic situation, you must ask:

1.   What is your company's mission?
2.   What is changing in the business climate?
3.   Where is your best possible market?
4.   What is your role as a sales manager?
5.   How clearly have you defined the sales process?

Sales management begins with your company's mission.
To organize salespeople, you must give them a higher purpose that ties them with their customers.
This mission must justify personal sacrifice.
It must offer a brighter future.
Dedication to this mission must inspire courage and honesty.

[19]Next, you have the climate.
It can be sunny or overcast.
It can be hot or cold.
It includes the timing of the seasons.

[23]Next is the terrain.
It can be distant or near.
It can be difficult or easy.
It can be open or narrow.
It also determines your life or death.

[28]Next is the commander.
He must be smart, trustworthy, caring, brave, and strict.

[30]Finally, you have your military methods.
They include the shape of your organization.
This comes from your management philosophy.
You must master their use.

[34]All five of these factors are critical.
As a commander, you must pay attention to them.
Understanding them brings victory.
Ignoring them means defeat.

Next, you must consider changes in the business climate.
These changes can be obvious or hidden.
They can speed up sales or slow sales down.
The changing climate includes the seasonal nature of business.

Next is your customer focus.
Do your salespeople need to travel or can they sell locally?
Which prospects are easier to sell to and which are harder?
Is your company's market wide open or a narrow niche?
Choosing the right marketplace determines success or failure.

Next is your role as a sales manager.
You must be knowledgeable, fair, sensitive, aggressive, and tough.

Finally, you need to have a clearly defined sales process.
What is the best way to organize your sales team?
Do you sell your company rather than just its products?
You must make selling easy.

All five factors are critical in analyzing your sales position.
You must be continuously evaluating your changing position.
Your personal knowledge is the key to your salespeople's success.
Overlooking any of these areas will lead to failure.

You must learn through planning. 2
You must question the situation.

³You must ask:
Which government has the right philosophy?
Which commander has the skill?
Which season and place has the advantage?
Which method of command works?
Which group of forces has the strength?
Which officers and men have the training?
Which rewards and punishments make sense?
This tells when you will win and when you will lose.
Some commanders perform this analysis.
If you use these commanders, you will win.
Keep them.
Some commanders ignore this analysis.
If you use these commanders, you will lose.
Get rid of them.

Plan an advantage by listening. 3
Adjust to the situation.
Get assistance from the outside.
Influence events.
Then planning can find opportunities and give you control.

2 You must continually expand your knowledge.
You must always ask your salespeople about specific situations.

You must question what you believe.
Does your higher mission resonate with your customers?
Are your making the right management decisions?
Which situations and markets give you an advantage?
What is the best way to organize the sales process?
Which salespeople are working together well?
Which salespeople really know what they are talking about?
How well are your sales incentive programs working?
You must know where your team is winning and losing sales.
Some of your salespeople are always doing their own analysis.
If you depend on these people, you will be successful.
Plan ways to hold on to them.
Some of your salespeople never analyze what they are doing.
If you rely on these people, you will fail.
Figure out ways to get rid of them.

3 You discover new opportunities by listening.
Adjust your sales strategy to your unique position.
Ask outsiders for suggestions on improving your sales process.
Make good things happen.
Your analysis uncovers the situations in which you are dominant.

Warfare is one thing. 4
It is a philosophy of deception.

3When you are ready, you try to appear incapacitated.
When active, you pretend inactivity.
When you are close to the enemy, you appear distant.
When far away, you pretend you are near.

7You can have an advantage and still entice an opponent.
You can be disorganized and still be decisive.
You can be ready and still be preparing.
You can be strong and still avoid battle.
You can be angry and still stop yourself.
You can humble yourself and still be confident.
You can be relaxed and still be working.
You can be close to an ally and still part ways.
You can attack a place without planning to do so.
You can leave a place without giving away your plan.

17You will find a place where you can win.
You cannot first signal your intentions.

4 Sales management requires one thing.
You must know how to control others' perception of reality.

When salespeople are overconfident, sow doubt.
When salespeople are worried, give them confidence.
When they think a sale is closing, suggest that it is far from closing.
When salespeople lack a sense of urgency, create it for them.

You can maintain authority and still woo your salespeople.
You can be confused and still appear decisive.
You can understand a sales situation and still play dumb.
You can control salespeople without fighting them.
You can be angry at salespeople and not let them know.
You can laugh at yourself and still maintain salespeople's respect.
You can seem confident when you are really scrambling.
You can like some salespeople and still get rid of them.
You can go after an opportunity that arises unexpectedly.
You can move away from certain customers without advertising it.

You know what your salespeople need to be successful.
Don't let them see how you are influencing them.

Manage to avoid battle until your organization can 5
count on certain victory.
You must calculate many advantages.
Before you go to battle, your organization's analysis can indi-
cate that you may not win.
You can count few advantages.
Many advantages add up to victory.
Few advantages add up to defeat.
How can you know your advantages without analyzing them?
We can see where we are by means of our observations.
We can foresee our victory or defeat by planning.

✦ ✦ ✦

5 You must focus your salespeople's efforts on customers and sales that they can certainly win.

You must see that you have an edge in those markets.

Before investing in a sales campaign, an objective analysis of your sales position can show that the odds are against you.

Your position doesn't give you an edge in certain markets.

Spending time on the best prospects adds up to sales success.

Investing in poor prospects adds up to poor sales.

How can you know where your opportunities lie without analysis?

As a sales manager, you have a broader view than your salespeople.

You must foresee their success or failure based on your analysis.

◆ ◆ ◆

Chapter 2

作戦

Going to War: The Key to Sales Profits

As a sales manager, one of your most important jobs is making sure
that your department's sales are profitable. Good strategy does not
define victory as simply winning sales. Strategy defines success
as making victory pay. This fact is fundamental, but many sales
managers forget that their job isn't simply meeting sales goals. It is
making profitable sales for their organization.

Running a sales department can be extremely expensive. The
potentially debilitating cost of selling destroys many organizations.
As a sales manager, you cannot spend your way to success. If you
waste your organization's limited resources, failure is certain. To
run a successful sales organization, you must run an efficient and
effective one. You can never assume that just because you invest
money your success will be certain.

Strategy teaches that the best way to run a profitable sales organization is to control costs. You cannot afford to hire more salespeople than you need. You cannot afford to take too long to close
sales. Local customers are always preferable to distant ones, simply
because they are more profitable. Sales are always more profitable
when the products and services they require are close to what your
company has provided before.

Solid companies are grown through profitable sales, not by
investment. Every sales effort must pay for itself. Rather than have
your company invest in growing your sales department, your profitable sales must pay for company growth.

Going to War

SUN TZU SAID:

Everything depends on your use of military philosophy. 1
Moving the army requires thousands of vehicles.
These vehicles must be loaded thousands of times.
The army must carry a huge supply of arms.
You need ten thousand acres of grain.
This results in internal and external shortages.
Any army consumes resources like an invader.
It uses up glue and paint for wood.
It requires armor for its vehicles.
People complain about the waste of a vast amount of metal.
It will set you back when you attempt to raise tens of thou-
sands of troops.

12Using a huge army makes war very expensive to win.
Long delays create a dull army and sharp defeats.
Attacking enemy cities drains your forces.
Long violent campaigns that exhaust the nation's resources
are wrong.

The Key to Sales Profits

1 Everything depends on your economic strategy.
Selling to distant customers requires constant travel.
This constant travel is a huge economic burden.
Your salespeople need computers and presentation tools.
You need money to compensate salespeople.
Your sales spending limits the spending in other company areas.
It is very easy to let spending for sales get out of control.
Sales spending limits marketing spending.
Sales spending limits product development.
Others in the organization complain about how much is spent on sales.
You are digging your company a grave when you build up a costly sales department.

A bloated sales department makes sales expensive to win.
Larger sales organizations react slowly and lose sales.
Targeting big, complex sales exhausts your sales force.
Long sales cycles that deplete your sales organization's resources are wrong.

[16]Manage a dull army.
You will suffer sharp defeats.
Drain your forces.
Your money will be used up.
Your rivals will multiply as your army collapses and they will
begin against you.
It doesn't matter how smart you are.
You cannot get ahead by taking losses!

[23]You hear of people going to war too quickly.
Still, you won't see a skilled war that lasts a long time.

[25]You can fight a war for a long time or you can make your
nation strong.
You can't do both.

Make no assumptions about all the dangers in using 2
military force.
Then you won't make assumptions about the benefits of
using arms either.

[3]You want to make good use of war.
Do not raise troops repeatedly.
Do not carry too many supplies.
Choose to be useful to your nation.
Feed off the enemy.
Make your army carry only the provisions it needs.

Your sales organization can react slowly.
You will then lose sales you should win.
Your sales team can take too long to close sales.
You will then invest too much in each sale.
Your competitors will then multiply and begin to target your company's customer base.
It doesn't matter how smart you think you are.
You cannot buy sales success by wasting resources.

You might hear of salespeople who move too quickly.
But no successful salesperson delays in closing sales.

You can let your salespeople take their time in closing sales or you can make your company successful.
You can't have it both ways.

2 Make no assumptions about the uncertainties of investing in a given sales cycle.
Then you won't overestimate the profits that your sales department will generate.

You want to make good use of your sales team.
Do not keep hiring more and more salespeople.
Do not burden the organization with a big payroll.
Focus on how you can support your company.
Generate sales growth from your sales.
Invest only in the sales team that your company needs.

The nation impoverishes itself shipping to troops that 3
are far away.
Distant transportation is costly for hundreds of families.
Buying goods with the army nearby is also expensive.
High prices also exhaust wealth.
If you exhaust your wealth, you then quickly hollow out your
military.
Military forces consume a nation's wealth entirely.
War leaves households in the former heart of the nation
with nothing.

[8]War destroys hundreds of families.
Out of every ten families, war leaves only seven.
War empties the government's storehouses.
Broken armies will get rid of their horses.
They will throw down their armor, helmets, and arrows.
They will lose their swords and shields.
They will leave their wagons without oxen.
War will consume 60 percent of everything you have.

Because of this, it is the intelligent commander's duty 4
to feed off the enemy.

[2]Use a cup of the enemy's food.
It is worth twenty of your own.
Win a bushel of the enemy's feed.
It is worth twenty of your own.

3 A company can invest too heavily in sending its salespeople to distant markets.

Big travel budgets are too costly for many companies.

Spending money on your sales efforts raises your prices.

High prices cost you sales.

If you waste resources, you have to eventually sacrifice needed parts of your company.

Expensive sales efforts can eat away at any company's strengths.

As the sales manager, you must protect your company's ability to compete.

Running out of money puts companies out of business.

Seven out of ten new businesses fail within two years.

Sales efforts can consume your company's needed reserves.

Failing sales teams must give up even needed travel.

They must abandon their sales and marketing tools.

They cannot attack competitors or defend their territories.

They must abandon valuable customers for lack of resources.

Poorly run sales teams can actually shrink your organization.

4 Because of this, your duty as a sales manager is to feed off of customers, not your own organization.

Pay your salespeople commissions based the sales they win.

This is twenty times better than paying them big salaries.

Steal away your competitors' best customers.

This is twenty times better than relying on existing customers.

⁶You can kill the enemy and frustrate him as well.
Take the enemy's strength from him by stealing away his
money.

⁸Fight for the enemy's supply wagons.
Capture his supplies by using overwhelming force.
Reward the first who capture them.
Then change their banners and flags.
Mix them in with your own wagons to increase your supply
line.
Keep your soldiers strong by providing for them.
This is what it means to beat the enemy while you grow
more powerful.

Make victory in war pay for itself. 5
Avoid expensive, long campaigns.
The military commander's knowledge is the key.
It determines if the civilian officials can govern.
It determines if the nation's households are peaceful or a
danger to the state.

♦ ♦ ♦

When you beat your competitors, you damage their credibility. By winning sales from competitors you steal away their ability to compete in the future.

Fight for competitors' most profitable customers.
Win these customers by focusing your salespeople on them.
Generously reward salespeople who win profitable sales.
Promote your services to your most profitable markets.
Increase your percentage of profitable sales in your base of recurring sales.
Keep your sales department strong by making it profitable.
This is what it means to weaken the opposition by helping your company grow more successful.

5 Your sales efforts are successful when they pay for themselves. Avoid a costly, bloated sales department.
Your knowledge as the sales manager is the key.
It enables your company's other divisions to invest in productivity.
It determines the happiness or frustration of your company's other employees.

◆ ◆ ◆

Chapter 3

謀攻

Planning an Attack: Your Sales Campaign's Focus

To successfully advance your company's sales position, you must define a clear focus for your sales compaign. Strategy teaches that sales strength doesn't come from size. Instead, it comes from unity and focus. To be successful as a sales manager, you must unite both your sales team and your organization as a whole. You must also focus on a campaign for sales growth that puts your competitors at a disadvantage.

Strategy teaches that unity and focus are required at every level of an organization. The more focused your sales organization is, the more efficiently it will work. The more divisions, distractions, and confusion in your sales efforts, the less successful your sales department will be.

When you are moving into new competitive areas, strategy teaches that you must avoid tackling well-developed competition. Instead, you must focus your sales campaign on areas where your competitors have yet to effectively organize their efforts.

Strategy teaches an incremental approach to success. You best competitors in small, focused engagements in which you have a clear advantage. It is your job as a sales manager to identify your company's relative strength in each sales situation and choose the right tactics.

Five types of managment knowledge determine your ability to unite and focus your sales team. You cannot afford to miscalculate the relative strength of your sales team in facing competitors.

Planning an Attack

SUN TZU SAID:

Everyone relies on the arts of war. 1
A united nation is strong.
A divided nation is weak.
A united army is strong.
A divided army is weak.
A united force is strong.
A divided force is weak.
United men are strong.
Divided men are weak.
A united unit is strong.
A divided unit is weak.

[12]Unity works because it enables you to win every battle you
fight.
Still, this is the foolish goal of a weak leader.
Avoid battle and make the enemy's men surrender.
This is the right goal for a superior leader.

The best policy is to attack while the enemy is still planning. 2
The next best is to disrupt alliances.
The next best is to attack the opposing army.
The worst is to attack the enemy's cities.

Your Sales Campaign's Focus

1 Everything depends on your sales management approach.
A united company is powerful.
A divided company is powerless.
A consistent sales message is trusted.
An inconsistent sales message is distrusted.
A focused sales campaign is successful.
An unfocused sales campaign is unsuccessful.
An organized sales department is strong.
A disorganized sales department is weak.
A united sales team works well together.
A divided sales team works poorly together.

Unity and focus work because they allow you to win every sales
challenge from the competition.
Still, this is the foolish goal of second-rate sales managers.
Discourage the competition from wanting to compete with you.
This is the best goal for the best sales managers.

2 The best campaign moves into markets before competitors.
The next best campaign divides your competitors' markets.
A worse campaign challenges your competitors' salespeople directly.
The worst campaign targets your competitors' strengths.

5This is what happens when you attack a city.
You can attempt it, but you can't finish it.
First you must make siege engines.
You need the right equipment and machinery.
It takes three months and still you cannot win.
Then you try to encircle the area.
You use three more months without making progress.
Your command still doesn't succeed and this angers you.
You then try to swarm the city.
This kills a third of your officers and men.
You are still unable to draw the enemy out of the city.
This attack is a disaster.

Make good use of war. 3
Make the enemy's troops surrender.
You can do this fighting only minor battles.
You can draw their men out of their cities.
You can do it with small attacks.
You can destroy the men of a nation.
You must keep your campaign short.

8You must use total war, fighting with everything you have.
Never stop fighting when at war.
You can gain complete advantage.
To do this, you must plan your strategy of attack.

What happens when you focus on your competitors' strengths?
You attempt to duplicate their abilities, but you never will.
First, you must build an organization that challenges theirs.
You must duplicate their systems and methods.
This takes time and you are always behind the competition.
You can try to keep competitors from getting ahead of you.
This wastes time by tying down your own sales organization.
Your attempts to copy competitors fail and this upsets you.
You then bet everything on a single sales effort.
This consumes your limited sales resources.
You cannot beat your competitors on their strengths.
This type of selling is a disaster.

3 A good sales campaign comes from good sales management.
Find an area where competitors do not want to compete.
You do this by identifying customers they see as unimportant.
You must pull their sales force away from established strengths.
You do this by winning away unimportant customers.
You want to discourage their salespeople.
You must win away customers quickly.

You must be completely committed to your sales focus.
Never hold back when you identify the right direction.
The right focus gives your salespeople complete superiority.
You must know where to put your sales resources.

[12]The rules for making war are:
If you outnumber enemy forces ten to one, surround them.
If you outnumber them five to one, attack them.
If you outnumber them two to one, divide them.
If you are equal, then find an advantageous battle.
If you are fewer, defend against them.
If you are much weaker, evade them.

[19]Small forces are not powerful.
However, large forces cannot catch them.

You must master command. 4
The nation must support you.

[3]Supporting the military makes the nation powerful.
Not supporting the military makes the nation weak.

[5]The army's position is made more difficult by politicians in
three different ways.
Ignorant of the whole army's inability to advance, they order
an advance.
Ignorant of the whole army's inability to withdraw, they
order a withdrawal.
We call this tying up the army.
Politicians don't understand the army's business.
Still, they think they can run an army.
This confuses the army's officers.

The rules for identifying the right sales campaign are simple.
A perfect sales campaign makes competitors give up the sale.
A great sales campaign pushes out competitors.
A good sales campaign pits competitors against one another.
A mediocre sales campaign wins when the situation is right.
A poor sales campaign is forced to defend existing customers.
A terrible sales campaign is forced to surrender customers.

Smaller companies cannot beat larger competitors directly.
However, large companies cannot keep up with small competitors.

4 You must focus your sales campaign.
Your company must support you.

Supporting its sales department makes any company strong.
Failing to support its sales department makes any company weaker.

There are three ways that your company's management can undermine your sales campaign's focus.
Ignorant about which type of sales your sales force cannot win, they want you to go after those sales.
Ignorant of which type of sales your sales force can win, they want you to abandon those sales.
This behavior constrains the sales force.
Company executives don't necessarily understand sales.
Still, they think that they can direct the sales force.
This undermines your best salespeople.

[12] Politicians don't know the army's chain of command.
They give the army too much freedom.
This will create distrust among the army's officers.

[15] The entire army becomes confused and distrusting.
This invites invasion by many different rivals.
We say correctly that disorder in an army kills victory.

You must know five things to win: 5
Victory comes from knowing when to attack and when to avoid battle.
Victory comes from correctly using both large and small forces.
Victory comes from everyone sharing the same goals.
Victory comes from finding opportunities in problems.
Victory comes from having a capable commander and the government leaving him alone.
You must know these five things.
You then know the theory of victory.

We say: 6
"Know yourself and know your enemy.
You will be safe in every battle.
You may know yourself but not know the enemy.
You will then lose one battle for every one you win.
You may not know yourself or the enemy.
You will then lose every battle."

♦ ♦ ♦

Management doesn't understand the need to prioritize sales.
It gives the sales force too much freedom.
This creates uncertainty in your sales leaders.

Your entire sales force can become confused and uncertain.
This invites competitors to win away your customers.
We know for certain that a lack of focus in sales leads to failure.

5 You must know five things to have a successful sales campaign.
You must know which prospects to target and which prospects to avoid.
You must know which prospects deserve a lot of resources and which require fewer resources.
Your sales team must share the same vision.
You must know how to turn problems into opportunities.
You must know how to control your sales force, and other executives must respect your decisions.
Master these five categories of knowledge.
You then know the philosophy of winning sales.

6 Pay attention.
Know your sales team and your competitors.
If you do, you can then meet your sales goals.
You may know your sales team but not your competitors.
Then, for every sale you make, you will lose another.
You may know neither your sales team nor your competitors.
Then you will lose every sale.

♦ ♦ ♦

Chapter 4

形

Positioning: Protecting and Advancing

As a sales manager, you must do two jobs. You must maintain your company's existing base of business. You must also look for new business to win. You need a stable base of support, so you must assist your salespeople in defending your company's recurring sales. However, the best defense is a good offense, so you must help your salespeople advance your company's position by identifying the best new types of opportunities.

Your first responsibility as a sales manager is defending your company's existing market position—that is, your existing customers. These sales are always the most profitable. While doing this, you must continually study the broader market to identify new opportunities for advancing your company's position. If you cannot defend your existing position, you do not have a dependable base to expand from.

After you spot a new opportunity, you must be able to move into a new position to capture it from your competition. When you focus on a new market, you must make sure you have the resources to tackle it. Sun Tzu provides a simple formula for calculating whether or not you can succeed in winning a new market position. This is done by calculating how close the new market is to your current one and how many resources you have available to go after it.

At the end of this chapter, Sun Tzu touches briefly on how critical having the right positioning is in getting what you want out of your salespeople.

Positioning

SUN TZU SAID:

Learn from the history of successful battles. 1
Your first actions should deny victory to the enemy.
You pay attention to your enemy to find the way to win.
You alone can deny victory to the enemy.
Only your enemy can allow you to win.

[6]You must fight well.
You can prevent the enemy's victory.
You cannot win unless the enemy enables your victory.

[9]We say:
You see the opportunity for victory; you don't create it.

Protecting and Advancing

1 Do not forget the value of the customers you already have.
First make sure that your salespeople keep their existing customers.
Look for weaknesses in your competitors to find new opportunities.
Your diligence is required to defend your existing base of sales.
Changes in the broader marketplace create your opportunities.

Use a strategic perspective.
You can prevent competitors from winning your customers.
Your competitors must leave openings to allow you to win theirs.

This is strategy.
See an opportunity to capture market share; don't try to create it.

You are sometimes unable to win. 2
You must then defend.
You will eventually be able to win.
You must then attack.
Defend when you have insufficient strength.
Attack when you have a surplus of strength.

7You must defend yourself well.
Save your forces and dig in.
You must attack well.
Move your forces when you have a clear advantage.

11You must always protect yourself until you can completely
triumph.

Some may see how to win. 3
However, they cannot position their forces where they must.
This demonstrates limited ability.

4Some can struggle to a victory and the whole world may
praise their winning.
This also demonstrates a limited ability.

6Win as easily as picking up a fallen hair.
Don't use all of your forces.
See the time to move.
Don't try to find something clever.
Hear the clap of thunder.
Don't try to hear something subtle.

2 You are not always positioned to win new business.
You must then defend the business you have.
You will eventually be in a position to expand your business.
Then you must move into new markets.
Defend your business when you lack the resources to expand it.
Expand when you have more resources than you need to defend.

You must defend your current position aggressively.
Keep your salespeople deeply involved with existing customers.
You must expand aggressively.
Move to new markets when you see a clear opportunity.

You must protect your existing customer base until you can win
new customers.

3 Some sales managers can see new sales opportunities.
However, they cannot organize their sales resources to win them.
This demonstrates a limited ability.

Other sales managers win new customers by expensive efforts that
seem like great successes.
This also shows a limited ability.

The right new markets are easily won.
Avoid using all your sales resources.
Wait for the right opportunities.
Don't try to demonstrate how clever you are.
See obvious opportunities.
Don't imagine opportunities where you want them.

[12]Learn from the history of successful battles.
Victory goes to those who make winning easy.
A good battle is one that you will obviously win.
It doesn't take intelligence to win a reputation.
It doesn't take courage to achieve success.

[17]You must win your battles without effort.
Avoid difficult struggles.
Fight when your position must win.
You always win by preventing your defeat.

[21]You must engage only in winning battles.
Position yourself where you cannot lose.
Never waste an opportunity to defeat your enemy.

[24]You win a war by first assuring yourself of victory.
Only afterward do you look for a fight.
Outmaneuver the enemy before the first battle and then
fight to win.

Examine where your salespeople have been successful.
Identify which customers they have won quickly and easily.
Good prospects are prospects who can obviously be won.
Your salespeople don't have to be brilliant to win them.
Your salespeople don't have to be courageous to approach them.

Focus your salespeople on prospects who can be sold to easily.
Salespeople must avoid difficult sales cycles.
Give them prospects that they must certainly win.
You always succeed when your salespeople cannot lose.

Your salespeople should only work on sales they can win.
Give them the right resources so that they cannot lose those sales.
Do not let them waste the opportunities that they are given.

Advance market positions by preparing your sales force for them.
Only then do you turn them loose in the market.
First, give your salepeople the knowledge they need to be successful,
then demand that they win sales.

You must make good use of war. 4
Study military philosophy and the art of defense.
You can control your victory or defeat.

[4]This is the art of war:
"1.   Discuss the distances.
2.   Discuss your numbers.
3.   Discuss your calculations.
4.   Discuss your decisions.
5.   Discuss victory.

[10]The ground determines the distance.
The distance determines your numbers.
Your numbers determine your calculations.
Your calculations determine your decisions.
Your decisions determine your victory."

[15]Creating a winning war is like balancing a coin of gold
against a coin of silver.
Creating a losing war is like balancing a coin of silver
against a coin of gold.

Winning a battle is always a matter of people. 5
You pour them into battle like a flood of water pouring into
a deep gorge.
This is a matter of positioning.

♦ ♦ ♦

4 You must make good use of your salespeople.
Teach them a higher mission by selling to win customer loyalty.
You control your salespeople's success and failure.

There is a system to sales management:
1.   Discuss the effort needed to win a new market.
2.   Discuss your sales resources.
3.   Discuss your sales priorities.
4.   Discuss your sales focus.
5.   Discuss your sales goals.

Your position determines the effort needed to win a new market.
The amount of effort determines the sales resources required.
The sales resources available are determined by your sales priorities.
Your sales priorities determine your sales focus.
Your sales focus determines your sales goals.

Creating a successful sales team requires focusing on the best opportunities instead of those that are second-rate.
Creating a losing sales team means focusing on second-rate opportunities instead of your best opportunities.

5 Success in sales management is always a matter of people.
You must make your salespeople's success a matter of following the path of least resistance.
This means positioning them correctly.

♦ ♦ ♦

Chapter 5

势

Momentum: Exciting Your Sales Force

Every sales manager must keep his or her sales force excited. In this chapter, Sun Tzu focuses on the need for creativity in strategy. You must harness this creativity to make selling for you and your company interesting and challenging.

Most salespeople are frightened by competitors who are larger and more powerful. One of Sun Tzu's basic principles is that size doesn't matter. Creativity, momentum, and excitement all count for much more in creating a successful sales force than size does.

Strategically, momentum comes from alternating between proven processes and innovative processes. In creating excitement in a sales force, you need to have clear standards and then continually introduce innovations. Standards and innovation require each other. Without standards, you just have chaos. However, standards without innovation do not create excitement or momentum.

Where do sales managers discover exciting innovations? Your salespeople are naturally coming up with good ideas every day. Every day, your customers are also coming up with success stories and new ways to view the value of your products and services. A sales manager's job is to find out what his or her salespeople are doing new, what is working, and why. As a sales manager, you must be like a pollinating bee, bringing ideas from one salesperson to the others.

When you create a constant sense of innovation and success, you create excitement.

Momentum

SUN TZU SAID:

You control a large group the same as you control a few. **1**
You just divide their ranks correctly.
You fight a large army the same as you fight a small one.
You only need the right position and communication.
You may meet a large enemy army.
You must be able to sustain an enemy attack without being
defeated.
You must correctly use both surprise and direct action.
Your army's position must increase your strength.
Troops flanking an enemy can smash them like eggs.
You must correctly use both strength and weakness.

It is the same in all battles. **2**
You use a direct approach to engage the enemy.
You use surprise to win.

[4]You must use surprise for a successful invasion.
Surprise is as infinite as the weather and land.
Surprise is as inexhaustible as the flow of a river.

Exciting Your Sales Force

1 You manage large sales forces the same as you run small ones.
You must assign responsibilities correctly.
You overcome large competitors the same as you do small ones.
Salespeople must communicate the company's strengths correctly.
Large competitors are intimidating.
Your sales force must know that it can win sales from these competitors successfully.
Your sales force must be more nimble and disciplined to succeed.
Sales success comes from the unique attributes of your company.
Your salespeople must know how to outmaneuver competitors.
You must train your salespeople on their competition's limitations.

2 All sales contests are the same.
Approach customers with products that they already understand.
Win sales by giving customers something new and exciting.

You must offer something new and exciting to win new markets.
Climate and customer changes make constant innovation possible.
Your salespeople's creativity must harness the power of change.

7You can be stopped and yet recover the initiative.
You must use your days and months correctly.

9If you are defeated, you can recover.
You must use the four seasons correctly.

11There are only a few notes in the scale.
Yet you can always rearrange them.
You can never hear every song of victory.

14There are only a few basic colors.
Yet you can always mix them.
You can never see all the shades of victory.

17There are only a few flavors.
Yet you can always blend them.
You can never taste all the flavors of victory.

20You fight with momentum.
There are only a few types of surprises and direct actions.
Yet you can always vary the ones you use.
There is no limit to the ways you can win.

24Surprise and direct action give birth to each other.
They are like a circle without end.
You cannot exhaust all their possible combinations!

Surging water flows together rapidly. 3
Its pressure washes away boulders.
This is momentum.

When you encourage creativity, every setback becomes temporary.
You can create excitement every day.

Salespeople find ways to win after a setback.
Your job is to make sure every salesperson learns new approaches.

Every sales presentation expresses only a handful of ideas.
Yet your salespeople can continually improve their presentations.
Your salespeople must all share each other's best work.

Every customer has only a few central concerns.
Train your salespeople to tailor every presentation to the customer.
Each sales presentation must be as unique as its target customer.

There are only a few forms of value.
Yet value changes from customer to customer, moment to moment.
Salespeople discover new product benefits every day.

Creativity and change excite your sales force.
Reward salespeople for using creativity and abiding by standards.
Recognize salespeople who come up with new approaches.
Constant innovation makes a sales force feel unstoppable.

Creativity without standards is just chaos.
Standards lead to innovation, which evolves into new standards.
You can improve anything through the process of innovation.

3 The pace of business change is increasing.
It is the pressure of change that washes away resistance.
Change gives your sales force its drive.

⁴A hawk suddenly strikes a bird.
Its contact alone kills the prey.
This is timing.

⁷You must fight only winning battles.
Your momentum must be overwhelming.
Your timing must be exact.

¹⁰Your momentum is like the tension of a bent crossbow.
Your timing is like the pulling of a trigger.

War is very complicated and confusing. 4
Battle is chaotic.
Nevertheless, you must not allow chaos.

⁴War is very sloppy and messy.
Positions turn around.
Nevertheless, you must never be defeated.

⁷Chaos gives birth to control.
Fear gives birth to courage.
Weakness gives birth to strength.

¹⁰You must control chaos.
This depends on your planning.
Your men must brave their fears.
This depends on their momentum.

¹⁴You have strengths and weaknesses.
These come from your position.

Your salespeople must quickly hit on your customers' current needs.
Sales excitement alone will win the sale.
Give people a reason to buy now.

Your salespeople must pursue only sales they can win.
Your company's momentum must give them confidence.
They must feel that they can close immediately.

Building excitement increases pressure on your salespeople.
Closing sales is what releases that pressure.

4 Sales reporting is often more fiction than fact.
Sales are unpredictable.
The sales manager's job is to create a sense of order.

Sales processes never go as planned.
Optimism changes to pessimism and back again.
Nevertheless, your salespeople cannot give up.

Salespeople's sloppiness demands your strictness.
Their apprehension demands your confidence.
Salespeople's shortcomings demand your management skills.

Salespeople want a sense of order.
You must enforce procedures that they can use.
Salespeople are easily unnerved.
You must keep their excitement high.

Your company has both strengths and weaknesses.
Yet you chose the company you work for.

¹⁶You must force the enemy to move to your advantage.
Use your position.
The enemy must follow you.
Surrender a position.
The enemy must take it.
You can offer an advantage to move him.
You can use your men to move him.
You can use your strength to hold him.

You want a successful battle. 5
To do this, you must seek momentum.
Do not just demand a good fight from your people.
You must pick good people and then give them momentum.

⁵You must create momentum.
You create it with your men during battle.
This is comparable to rolling trees and stones.
Trees and stones roll because of their shape and weight.
Offer men safety and they will stay calm.
Endanger them and they will act.
Give them a place and they will hold.
Round them up and they will march.

¹³You make your men powerful in battle with momentum.
This should be like rolling round stones down over a high,
steep cliff.
Momentum is critical.

✦ ✦ ✦

You must keep salespeople excited about their opportunities.

Share your views of the company with them.

You must put the competition into perspective.

Be honest with your salespeople about a competitor's strengths.

Your salespeople must agree with your assessment.

Explain why your company has the advantage over the competition.

Your salespeople themselves must be part of that reason.

Your company's strengths must keep the customers that they win.

5 You must beat your competitors.

Your success depends upon your sales force's excitement.

You cannot boss your sales force into meeting sales quota.

You keep them excited about the company and sales will follow.

Creating excitement is hard work.

You must use ideas that your salespeople come up with every day.

You must use ideas that the salespeople themselves like.

Salespeople like ideas when customers easily accept them.

Your salespeople must feel secure to be confident.

Your salespeople must feel pressure to get the work done.

Clear procedures provide methods salespeople know will work.

Clear goals create the pressure your people need to stay motivated.

You make your sales force powerful by keeping it excited.

You must use your salespeople's natural tendency to copy each other's past successes.

Excitement is the key.

♦ ♦ ♦

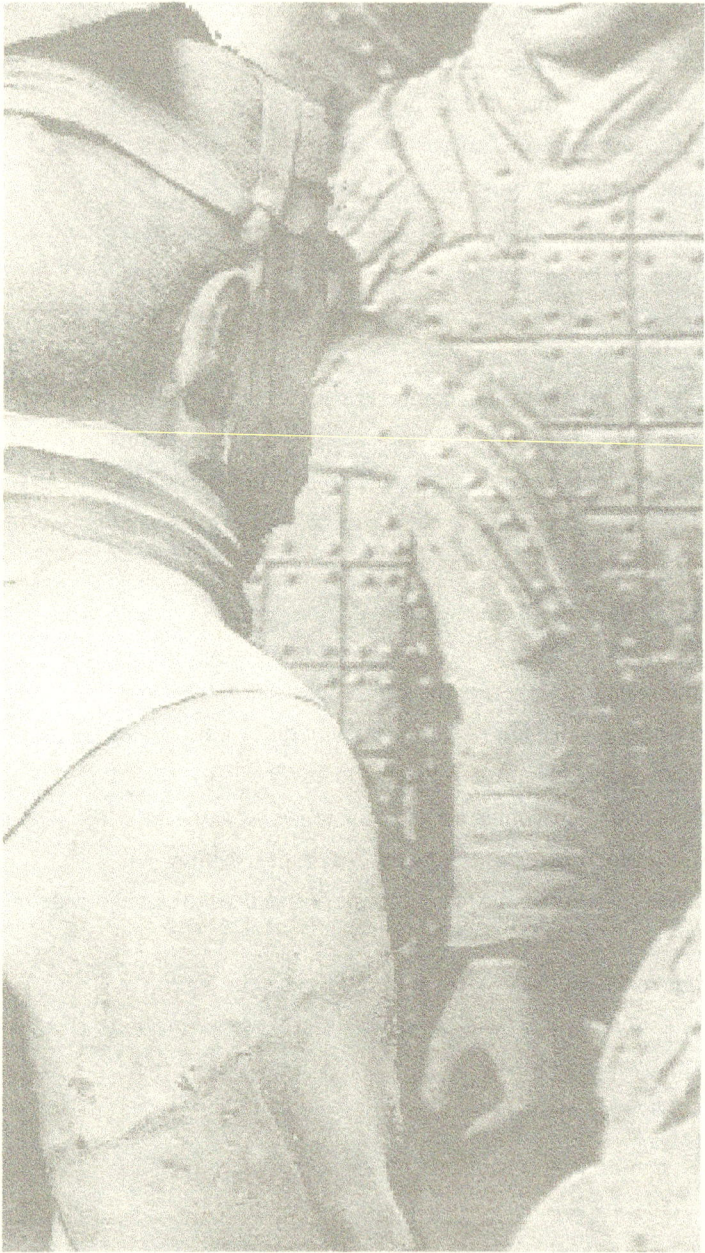

Chapter 6

虛實

Weakness and Strength: Expansion Opportunities

Sales managers work on the front lines of business competition. Sales managers must design sales campaigns, sales processes, and sales offers that put the competition at a disadvantage. Controlling salespeople is easy if you identify expansion opportunities that put competitors on the defensive.

Philosophically, Sun Tzu's system works when you balance your strengths against your competition's weaknesses. The easiest way to do this is by using speed. You want to get to your customers with new ideas long before your competition does. Taking the initiative in the market is the most important aspect of expanding your business.

No matter how good sales are today, it is always your job to make them better tomorrow. This means that you must identify methods by which you can expand sales. There are only two ways to do this: discover new customers who need your product or discover new needs for your product among existing customers.

The market is always changing. New products and new needs create expansion opportunities. As a sales manager, you must identify problems and the opportunities they create and reassign your sales resources appropriately.

Good sales managers are flexible in constantly adjusting to the opportunities of the marketplace. If you and your company can react more quickly than your competitors, you will maintain the initiative and your sales force will be successful.

Weakness and Strength

SUN TZU SAID:

Always arrive first to the empty battlefield to await the 1
enemy at your leisure.
After the battleground is occupied and you hurry to it, fighting is more difficult.

³You want a successful battle.
Move your men, but not into opposing forces.

⁵You can make enemies come to you.
Offer them an advantage.
You can make enemies avoid coming to you.
Threaten them with danger.

⁹When enemies are fresh, you can tire them.
When they are well fed, you can starve them.
When they are relaxed, you can move them.

Expansion Opportunities

1 As a sales manager, you must always stay several steps ahead of
your competitors.
If your competitors have the initiative, you will have problems
growing your sales.

You want all your salespeople to be successful.
You want them to create new sales, not take sales from one another.

You may want to encourage competitors to follow your lead.
Make offers that they can match.
You can stop your competitors from copying your offers.
Make offers that competitors cannot match.

If your competitors are too comfortable, challenge them.
If your competitors are self-satisfied, steal their customers.
If your competitors have grown lazy, push them out of markets.

Leave any place without haste. 2
Hurry to where you are unexpected.
You can easily march hundreds of miles without tiring.
To do so, travel through areas that are deserted.
You must take whatever you attack.
Attack when there is no defense.
You must have walls to defend.
Defend where it is impossible to attack.

9Be skilled in attacking.
Give the enemy no idea where to defend.

11Be skillful in your defense.
Give the enemy no idea where to attack.

Be subtle! Be subtle! 3
Arrive without any clear formation.
Ghostly! Ghostly!
Arrive without a sound.
You must use all your skill to control enemies' decisions.

6Advance where they can't defend.
Charge through their openings.
Withdraw where enemies cannot chase you.
Move quickly so that they cannot catch you.

2 Good sales managers are slow to abandon existing business.
Great sales managers are quick to embrace new opportunities.
You can dramatically change what, how, and to whom you sell.
You must identify needs that are being left unsatisfied.
New opportunities must prove themselves quickly.
Find product niches where there are no alternatives.
Protect every customer and market your salespeople win.
Don't leave competitors an opening to steal them away.

As a sales manager, your job is to grow your business.
Uncover growth areas that are being overlooked.

As a sales manager, your job is to protect your business.
Stop your salespeople from creating easy targets for competitors.

3 Sales managers need to be sensitive to changes.
Observe the competition with an open mind.
Monitor your marketplace inconspicuously.
Show up where you are least expected.
You can control how competitors view a marketplace.

Expand your business where competitors are having problems.
Make your salespeople respond quickly.
Move salespeople out of areas where competitors are hurting them.
React quickly before your salespeople lose heart.

[10]Always pick your own battles.
Enemies can hide behind high walls and deep trenches.
Do not try to win by fighting them directly.
Instead, attack a place that they must recapture.
Avoid the battles that you don't want.
You can divide the ground and yet defend it.
Don't give enemies anything to win.
Divert them by coming to where you defend.

Make other men take a position while you take none. 4
Then focus your forces where enemies divide their forces.
Where you focus, you unite your forces.
When enemies divide, they create many small groups.
You want your large group to attack one of their small ones.
Then you have many men where enemies have but a few.
Your larger force can overwhelm a small one.
Then go on to the next small enemy group.
You can take them one at a time

You must keep the place that you have chosen as a 5
battleground a secret.
Enemies must not know it.
Force enemies to prepare their defense in many places.
You want enemies to defend many places.
Then you can choose where to fight.
Their forces will be weak there.

You decide where to focus your sales force.

Opponents can protect their existing customers extremely well.

Do not directly challenge them.

Instead, go after a new market they will eventually need.

Avoid battles you cannot win.

Find segments of competitors' markets where you have an advantage.

Focus on those niches that favor your products.

Go after their markets so they don't come after yours.

4 Let competitors make choices while you leave your options open.

Identify key niches that your competitors are overlooking.

Adjust territories to focus salespeople on these opportunities.

Let your competitors divide their resources among too many areas.

Overwhelm competitors by using their weaknesses against them.

Concentrate your sales force in markets where competitors are weak.

Your market commitment highlights the competition's indifference.

Win one niche, then go on to the next.

You grow sales one market segment at a time.

5 Train your salespeople not to talk to competitors about your sales focus.

Your competition will misinterpret what you are doing.

Never make the mistake of spreading your resources too thin.

You cannot be the best at everything.

Your salespeople need a clear target.

Ask your sales force who competitors are neglecting.

7If they reinforce their front lines, they deplete their rear.
If they reinforce their rear, they deplete their front.
If they reinforce their right flank, they deplete their left.
If they reinforce their left flank, they deplete their right.
Without knowing the place of attack, they cannot prepare.
Without knowing the right place, they will be weak everywhere.

13Your enemies have weak points.
Prepare your men against them.
Your enemies have their strong points.
Make their men prepare themselves against you.

You must know the battleground. 6
You must know the time of battle.
You can then travel a thousand miles and still win the battle.

4The enemy should not know the battleground.
He shouldn't know the time of battle.
His left flank will be unable to support his right.
His right will be unable to support his left.
His front lines will be unable to support his rear.
His rear will be unable to support his front.
His support is distant even if it is only ten miles away.
What unknown place can be close?

12You control the balance of forces.
The enemy may have many men but they are superfluous.
How can they help him to victory?

Watch your competitors as they shift their priorities.
Every change creates new opportunities for you.
Every shift in resources weakens existing positions.
A lack of clear priorities depletes every organization's credibility.
If you keep your focus a secret, your competitors won't figure it out.
Most will try to be all things to all potential markets.

Your competitors always have weak points.
Focus your sales force on attacking them.
Your competitors also have their strengths.
Stop your salespeople from attacking those strengths.

6 You must train your salespeople to know the target niche.
You must teach them the trends affecting customers.
Given the right knowledge, even difficult sales can be profitable.

Train your salespeople to keep customer information a secret.
Train your salespeople not to talk about your sales plans.
The less your competition knows, the weaker it is.
Encourage your salespeople to mislead competitors.
Competitors will naturally misinterpret your sales appeal.
Your salespeople should encourage that tendency.
Your competition may have potential in your niche.
Your competitors' ignorance is your best friend.

Information is the source of all selling power.
Your competitors may have much larger sales forces.
Ignorant salespeople are never a threat.

[15]We say:
You must let victory happen.

[17]The enemy may have many men.
You can still control him without a fight.

When you form your strategy, know the strengths and **7**
weaknesses of your plan.
When you execute a plan, know how to manage both action
and inaction.
When you take a position, know the deadly and the winning
grounds.
When you enter into battle, know when you have too many
or too few men.

[5]Use your position as your war's centerpiece.
Arrive at the battle without a formation.
Don't take a position in advance.
Then even the best spies can't report it.
Even the wisest general cannot plan to counter you.
Take a position where you can triumph using superior numbers.
Keep opposing forces ignorant.
Everyone should learn your location after your position has
given you success.
No one should know how your location gives you a winning
position.
Make a successful battle one from which the enemy cannot
recover.
You must continually adjust your position to his position.

Train your sales force.
Sales flow to those who are knowledgeable.

Your competitors can have many strengths.
They cannot fight if they lack key information.

7 When planning a sales campaign, identify both what can go right and what can go wrong.
During the campaign, assign what needs doing and stop people from wasting their time.
When targeting a market segment, know which customers are valuable and which are worthless.
You must design all campaigns and pick markets to fit the size of your sales force.

You must build your campaign on your company's past success.
Every campaign must be opportunistic.
Let the changing situation dictate your policies.
Keep competitors in the dark as long as possible.
Good competitors can always react if you give them time.
Your campaign must focus where the competition is weak.
Train your salespeople to mislead competitors.
Competitors should identify your target market after your salespeople are entrenched.
Competitors should appreciate the appeal of your proposals only after it is too late to counter them.
Competitors can only blunt your campaign if you give them the time to respond.
Adjust to competitors' moves faster than they can adjust to yours.

Manage your military position like water. 8
Water takes every shape.
It avoids the high and moves to the low.
Your war can take any shape.
It must avoid the strong and strike the weak.
Water follows the shape of the land that directs its flow.
Your forces follow the enemy who determines how you win.

[8]Make war without a standard approach.
Water has no consistent shape.
If you follow the enemy's shifts and changes, you can always
find a way to win.
We call this shadowing.

[12]Fight five different campaigns without a firm rule for victory.
Use all four seasons without a consistent position.
Your timing must be sudden.
A few weeks determine your failure or success.

♦ ♦ ♦

8 Good sales managers are flexible.
Salespeople must be trained to adapt.
Keep them out of difficult sales cycles and working on easy ones.
Good sales can take any number of forms.
Good sales avoid what is difficult and offer what is easy.
Good sales managers let the situation shape their offering.
If you follow your customers' lead, you will design good programs.

You must avoid being too rigid in your policies.
Consistent success requires constant adjustment.
If you equal your competition's offers quickly, you maintain the initiative.
Your salespeople must monitor them.

As the sales manager, you can adjust the rules to fit the situation.
In a year, everything in your environment can change.
Adjust to shifts instantly.
You can fall behind very quickly if you don't.

◆ ◆ ◆

Chapter 7

軍 爭

Armed Conflict: Company Politics

Sales managers are often caught in the crossfire of company politics. Selling affects every part of the organization. To be a good sales manager, it is not enough just to run the sales department well. Your success depends upon maintaining a good working relationship with every part of the organization. As every sales manager knows, this is always a challenge given the separate responsibilities of every division.

Political struggles within a company are always dangerous, but they are especially dangerous for sales managers. Sales managers who make themselves a target will not survive.

Your success depends on meeting sales goals. Internal political battles can only distract from your efforts to meet those goals. The more resources that you devote to internal politics, the less successful your sales department will be.

Most conflict within organizations can be controlled by controlling other people's perceptions. Much of working with the rest of the organization comes down to good communication skills, not only on your part, but on the part of all your salespeople. To make your job easier, you should encourage the sales department to develop friendships throughout the organization.

The more excited your salespeople are about the company, the more successful they will be. Maintain a positive attitude in your sales department about the company and the rest of the company will have a positive attitude about your sales department.

Armed Conflict

<small>SUN TZU SAID:</small>

Everyone uses the arts of war. 1
You accept orders from the government.
Then you assemble your army.
You organize your men and build camps.
You must avoid disasters from armed conflict.

6Seeking armed conflict can be disastrous.
Because of this, a detour can be the shortest path.
Because of this, problems can become opportunities.

9Use an indirect route as your highway.
Use the search for advantage to guide you.
When you fall behind, you must catch up.
When you get ahead, you must wait.
You must know the detour that most directly accomplishes
your plan.

14Undertake armed conflict when you have an advantage.
Seeking armed conflict for its own sake is dangerous.

Company Politics

1 Every department in your organization is jockeying for position. You each have your own responsibilities in the company. You are personally responsible for the sales force. You create the sales systems and enforce sales policies. You must avoid conflict with other departments.

You cannot get ahead by playing company politics. You must find ways around interdepartmental problems. You must think of those problems as opportunities to improve.

You must be willing to go out of your way to make progress. You can always find a way to improve any situation. Change sales procedures when they create bottlenecks. When other departments are bottlenecks, you must be patient. You must find an indirect way of getting those departments to improve the situation.

Use company politics when it creates long-term improvement. Playing politics just to win internal battles is fatal.

You can build up an army to fight for an advantage. 2
Then you won't catch the enemy.
You can force your army to go fight for an advantage.
Then you abandon your heavy supply wagons.

5You keep only your armor and hurry after the enemy.
You avoid stopping day or night.
You use many roads at the same time.
You go hundreds of miles to fight for an advantage.
Then the enemy catches your commanders and your army.
Your strong soldiers get there first.
Your weaker soldiers follow behind.
Using this approach, only one in ten will arrive.
You can try to go fifty miles to fight for an advantage.
Then your commanders and army will stumble.
Using this method, only half of your soldiers will make it.
You can try to go thirty miles to fight for an advantage.
Then only two out of three get there.

18If you make your army travel without good supply lines,
your army will die.
Without supplies and food, your army will die.
If you don't save the harvest, your army will die.

2 You can devote resources to fighting political battles.
Other department managers will just avoid you.
You can force your salespeople to get involved in internal battles.
You then forget that your real job is making sales.

You can defend yourself and try to prove your opponents wrong.
You can work day and night.
You can try to cover all your bases.
You can move mountains to discredit an internal rival.
Then your rival will focus on shortcomings in the sales department.
The great sales that you have made are forgotten.
Your sales errors are the focus of the discussion.
Only a small fraction of your efforts will be rewarded.
You can make more limited moves in fighting internal battles.
This will still get you into trouble.
You will be only half as successful as you could be.
You shouldn't go even a little out of your way to play politics.
It costs you more than it is worth.

Without real support from every department in the company, you
cannot make sales.
Without products and service, you cannot make sales.
If you do not conserve company resources, you cannot make sales.

²¹Do not let any of your potential enemies know what you
are planning.
Still, you must not hesitate to form alliances.
You must know the mountains and forests.
You must know where the obstructions are.
You must know where the marshes are.
If you don't, you cannot move the army.
If you don't, you must use local guides.
If you don't, you can't take advantage of the terrain.

You make war using a deceptive position. 3
If you use deception, then you can move.
Using deception, you can upset the enemy and change the
situation.
You can move as quickly as the wind.
You can rise like the forest.
You can invade and plunder like fire.
You can stay as motionless as a mountain.
You can be as mysterious as the fog.
You can strike like sounding thunder.

¹⁰Divide your troops to plunder the villages.
When on open ground, dividing is an advantage.
Don't worry about organization; just move.
Be the first to find a new route that leads directly to a
winning plan.
This is how you are successful at armed conflict.

You must keep opponents from knowing how you plan to change sales procedures.

Always, you must look for allies in the organization.

You must know both top management and regular employees.

You must know where the potential roadblocks are.

You must know where you might get bogged down.

If you don't, you won't improve your sales department.

You must make friends in every department.

If you don't, you won't get promoted within the organization.

3 You must control appearances to succeed at company politics.
You can make progress only when you control appearances.

By changing policy and procedures, you can placate opponents and improve operations.

You can change policies quickly using a light touch.

You can gradually revolutionize systems over time.

You can win political capital by making big sales.

You can avoid unwanted attention by appearing solid.

You can move silently behind the scenes.

You can make noise to demand attention.

Divide your salespeople to nurture contacts in other departments.

Avoid internal conflict by making friends throughout the company.

Don't worry about upper management; work out problems quickly.

Your salespeople should always be looking for better ways of working with others in the company.

This is how you are successful at company politics.

Military experience says: 4
"You can speak, but you will not be heard.
You must use gongs and drums.
You cannot really see your forces just by looking.
You must use banners and flags."

6You must master gongs, drums, banners, and flags.
Place people as a single unit where they can all see and hear.
You must unite them as one.
Then the brave cannot advance alone.
The fearful cannot withdraw alone.
You must force them to act as a group.

12In night battles, you must use numerous fires and drums.
In day battles, you must use many banners and flags.
You must position your people to control what they see and
hear.

You control your army by controlling its morale. 5
As a general, you must be able to control emotions.

3In the morning, a person's energy is high.
During the day, it fades.
By evening, a person's thoughts turn to home.
You must use your troops wisely.
Avoid the enemy's high spirits.
Strike when his men are lazy and want to go home.
This is how you master energy.

4 Communication is the key to politics.
The success of your sales department will not speak for itself.
You must promote your sales successes to win capital.
You cannot know exactly what your salespeople are doing.
You must have good sales reporting systems.

Use reporting and promotion to get executive attention.
Make sure that your sales successes are known to everyone.
Celebrate sales successes to bring the company together.
Do not let the salespeople take credit alone.
Insist on sharing credit with everyone involved.
You want your salespeople to work with others.

The less dramatic your sales progress, the more visibility it needs.
Even when sales progress is obvious, you must still advertise it.
You must have channels of communication throughout the organi-
zation to control impressions.

5 You control your salespeople by keeping them excited.
As sales manager, you determine how they feel about the company.

When salespeople are newly hired, they are excited by the company.
As time goes on, their excitement fades.
When their excitement is gone, you have to get rid of them.
You must manage your salespeople well.
Don't criticize salespeople who are too enthusiastic.
Criticize salespeople when they get lazy and uninvolved.
This is how you maintain excitement in a sales force.

[10]Use discipline to await the chaos of battle.
Keep relaxed to await a crisis.
This is how you master emotion.

[13]Stay close to home to await a distant enemy.
Stay comfortable to await the weary enemy.
Stay well fed to await the hungry enemy.
This is how you master power.

Don't entice the enemy when his ranks are orderly. 6
You must not attack when his formations are solid
This is how you master adaptation.

[4]You must follow these military rules.
Do not take a position facing the high ground.
Do not oppose those with their backs to the wall.
Do not follow those who pretend to flee.
Do not attack the enemy's strongest men.
Do not swallow the enemy's bait.
Do not block an army that is heading home.
Leave an escape outlet for a surrounded army.
Do not press a desperate foe.
This is how you use military skills.

✦ ✦ ✦

Use sales discipline to avoid internal problems.
Keep calm in every crisis.
You must control everyone's emotions.

Keep in touch with sales and await critics who are out of touch.
Manage sales well and await critics who are poor managers.
Make profitable sales and await critics who are less profitable.
This is how you develop influence.

6 Do not attract criticism from the company's best managers.
You must avoid conflicts with those who are doing a good job.
You must be flexible enough to admit errors.

Sales management has its own rules:
Do not oppose the viewpoints of top executives.
Do not give criticism without offering solutions.
Do not copy people who have failed in the past.
Do not disparage your company's successful managers.
Do not believe everything you hear.
Do not get into battles with people who are about to retire.
Leave anyone you oppose a way to save face.
Do not push anyone too far.
This is the strategy of sales management.

✦ ✦ ✦

Chapter 8

九變

Adaptability: Crucial Decisions

Your organization depends upon you as a sales manager to make
certain crucial decisions. Only sales managers are in a position
to recognize emerging problems and address them quickly. This
chapter serves as an introduction to the next three chapters, which
address different types of situations and how to respond to them
correctly. No one other than the sales manager is positioned to
recognize these situation and choose the proper course of action.

As the pace of change in business increases, more and more
critical decisions fall to sales managers. As a sales manager, you are
the key decision-maker working on the front lines of your business. Successful strategies must be dynamic because the world is
always changing. If you do not understand how to address business
changes, you will not be successful as a sales manager.

No sales policy or procedure is sacred. Neither your methods
nor plans can be carved in stone. No matter how well these have
worked in the past, you must be willing to change them as the
evolving situation warrants.

You must take the initiative to throw your company's competitors off balance. Either you use changes in business to manipulate
your competitors or they will use those same changes to manipulate
you.

You are also responsible for identifying the character flaws in
your salespeople that may prevent them from adapting to the new,
fast-changing world of business.

Adaptability

Everyone uses the arts of war. 1
As a general, you get your orders from the government.
You gather your troops.
On dangerous ground, you must not camp.
Where the roads intersect, you must join your allies.
When an area is cut off, you must not delay in it.
When you are surrounded, you must scheme.
In a life-or-death situation, you must fight.
There are roads that you must not take.
There are armies that you must not fight.
There are strongholds that you must not attack.
There are positions that you must not defend.
There are government commands that must not be obeyed.

[14]Military leaders must be experts in knowing how to adapt
to find an advantage.
This will teach you the use of war.

[16]Some commanders are not good at making adjustments to
find an advantage.
They can know the shape of the terrain.
Still, they cannot find an advantageous position.

Crucial Decisions

1 All sales managers must think strategically.
As a sales manager, you get your sales goals from your company.
You hire your salespeople.
When business is difficult, sales managers cannot relax.
When sales require partners, sales managers make new alliances.
When markets go nowhere, sales managers don't stay in them.
When business is overwhelming, sales managers innovate.
When alternatives seem exhausted, sales managers don't give up.
You alone can choose paths that lead to more opportunities.
You alone can identify customers that you can easily win.
You alone can pick competitors who are vulnerable.
You alone can establish positions that are easy to defend.
You must do what your company needs, not just what people want.

Good sales managers know how to continually adjust their plans to create new opportunities.
Testing your ideas teaches you which approaches are best.

Some sales managers cannot change their viewpoint in order to see their opportunities.
You can know your business well.
You must also identify opportunities to improve your situation.

[19]Some military commanders do not know how to adjust
their methods.
They can find an advantageous position.
Still, they cannot use their men effectively.

You must be creative in your planning. 2
You must adapt to your opportunities and weaknesses.
You can use a variety of approaches and still have a
consistent result.
You must adjust to a variety of problems and consistently
solve them.

You can deter your potential enemy by using his 3
weaknesses against him.
You can keep your potential enemy's army busy by giving it
work to do.
You can rush your potential enemy by offering him an
advantageous position.

You must make use of war. 4
Do not trust that the enemy isn't coming.
Trust your readiness to meet him.
Do not trust that the enemy won't attack.
Rely only on your ability to pick a place that the enemy can't
attack.

Other sales managers can see new opportunities but they cannot adjust their sales policies.

You must do more than see opportunities.

Make it possible for salespeople to take advantage of them.

2 You must change your sales procedures.

Identify new shortcuts and address lingering problems.

The faster business changes, the more adjustments you must make to maintain consistency.

Every problem in sales management represents a new opportunity for improvement.

3 It is your responsibility as a sales manager to identify the key weaknesses in your competitors.

If you are doing your job correctly, your competitors must adjust to what your salespeople are doing.

If you leave the right bait tempting your competitors, you can lure them into difficult situations.

4 It is the sales manager's job to worry about competitors.

Ask your salespeople about what your competitors are doing.

Then you can keep ahead of the competition.

Ask your salespeople where your company's weaknesses are.

Then you can deal with those weaknesses before your competitors can exploit them.

You can exploit five different faults in a leader. 5
If he is willing to die, you can kill him.
If he wants to survive, you can capture him.
He may have a quick temper.
You can then provoke him with insults.
If he has a delicate sense of honor, you can disgrace him.
If he loves his people, you can create problems for him.
In every situation, look for these five weaknesses.
They are common faults in commanders.
They always lead to military disaster.

[11]To overturn an army, you must kill its general.
To do this, you must use these five weaknesses.
You must always look for them.

5 You must identify the character flaws in your salespeople.
If salespeople are too confident, they alienate customers.
If salespeople lack courage, they do not make calls.
If salespeople are temperamental, they waste time.
If salespeople are sensitive to rejection, they cannot close sales.
If salespeople cannot take criticism, they cannot be corrected.
If salespeople want to be popular, they give profits away.
In every discussion, look for signs of these problems.
A career in selling attracts people with these flaws.
A flawed sales force leads to disaster.

To build a successful sales department, you must survive.
You must be aware of your own weaknesses.
You must analyze them objectively.

✦ ✦ ✦

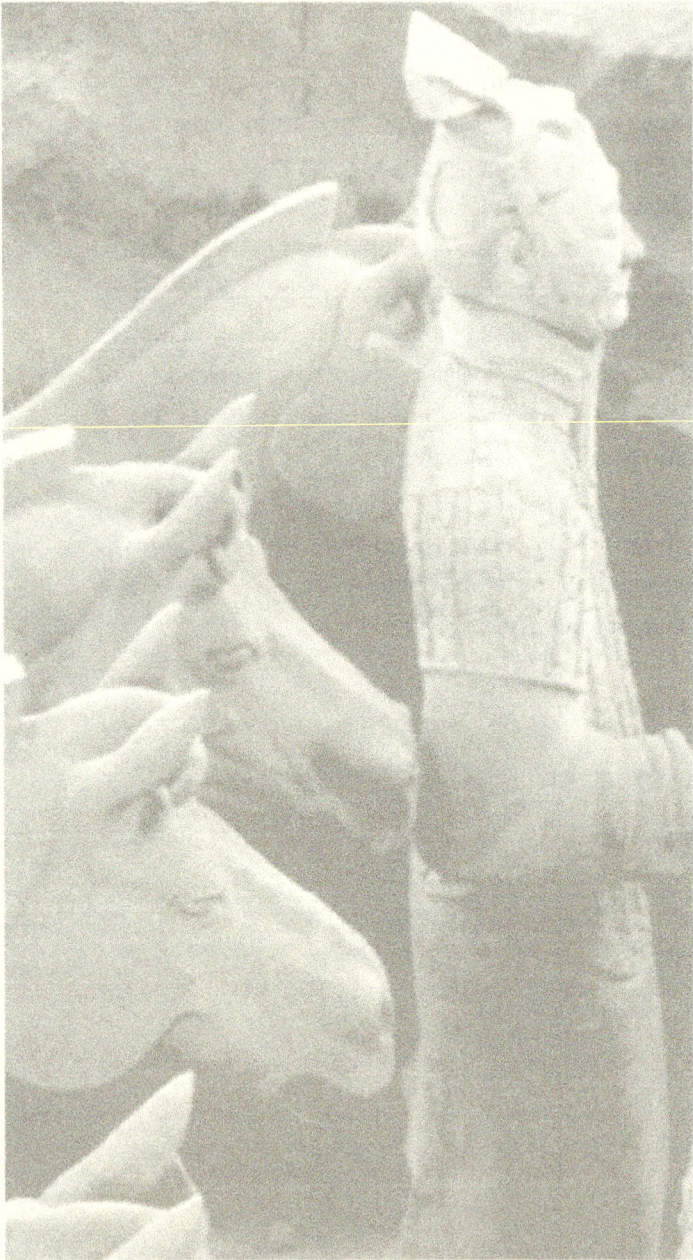

Chapter 9

行軍

Armed March: Outmaneuvering Competitors

As a sales manager, you must know how to teach your salespeople about beating competitors. Strategically, you don't want your salespeople to attack competitors. Instead, you want salespeople to outmaneuver competitors. This means working in a way that puts your competition at a constant disadvantage.

Strategy is a process that is always situation driven. Different sales situations demand different competitive responses on the part of your salespeople. Your salespeople must recognize the category of sales situation that they are in. They must also know how you want them to respond to it.

Strategically, good ethics are a critical aspect of meeting competitive threats. The pressure of competition can tempt salespeople into making foolish, short-sighted mistakes. As a sales manager, you must set a clear, ethical example for your sales team.

Every salesperson inevitably encounters immovable obstacles in the sales process. A sales manager must teach salespeople not to waste time on these obstacles but to go around them.

As the central information resources within a sales department, sales managers are best positioned to understand what is happening with competitors. Sun Tzu gives a long list of ways to interpret your competitors' situation based upon signs in the market.

In the end, outflanking competitors comes down to recruiting the right salespeople and training them correctly. Without enough trained people, you can not effectively counter the competition.

Armed March

SUN TZU SAID:

1 Anyone moving an army must adjust to the enemy. 1
When caught in the mountains, rely on their valleys.
Position yourself on the heights facing the sun.
To win your battles, never attack uphill.
This is how you position your army in the mountains.

6When water blocks you, keep far away from it.
Let the invader cross the river and wait for him.
Do not meet him in midstream.
Wait for him to get half his forces across and then take
advantage of the situation.

10You need to be able to fight.
You can't do that if you are caught in water when you meet
an invader.
Position yourself upstream, facing the sun.
Never face against the current.
Always position your army upstream when near the water.

Outmaneuvering Competitors

1 You manage a winning sales force by outmaneuvering competitors.
In large organizations, salespeople make progress in lower echelons.
Teach them to work up to higher, more visible decision-makers.
Your sales force cannot be successful battling upper management.
This is how your sales force must work within large organizations.

Keep your salespeople out of prospects that are reorganizing.
Let your competitors waste their time amid the chaos of change.
Your salespeople should not compete directly in these situations.
The more deeply mired your competitors become, the more oppor-
tunities there are elsewhere for your salespeople.

You have to defend your existing business.
Your salespeople must keep competitors out of existing accounts
that are reorganizing.
Train them to position your company for the future.
Teach them not to fight against a customer's transitions.
Your sales force must learn to navigate customer transitions.

¹⁵You may have to move across marshes.
Move through them quickly without stopping.
You may meet the enemy in the middle of a marsh.
You must keep on the water grasses.
Keep your back to a clump of trees.
This is how you position your army in a marsh.

²¹On a level plateau, take a position that you can change.
Keep the higher ground on your right and to the rear.
Keep danger in front of you and safety behind.
This is how you position yourself on a level plateau.

²⁵You can find an advantage in all four of these situations.
Learn from the great emperor who used positioning to
conquer his four rivals.

Armies are stronger on high ground and weaker on low. 2
They are better camping on sunny southern hillsides than
on shady northern ones.
Provide for your army's health and place men correctly.
Your army will be free from disease.
Done correctly, this means victory.

⁶You must sometimes defend on a hill or riverbank.
You must keep on the south side in the sun.
Keep the uphill slope at your right rear.

⁹This will give the advantage to your army.
It will always give you a position of strength.

Your salespeople may get into complicated sales situations.
You must train them to move through these sales processes quickly.
Train your salespeople to be competitive in confusing accounts.
They must establish a well-defined sales position.
You cannot let competitors push your salespeople into trouble.
Your salespeople must avoid missteps on uncertain ground.

In most accounts, your salespeople should keep their options open.
Encourage your salespeople to move up in an organization.
Keep an eye on competitors and prepare a fallback position.
This is how your sales force succeeds in typical accounts.

Your sales force must be prepared to win in every situation.
The best sales managers can diagnose every situation and instruct
their salespeople correctly.

2 A sales force with good ethics is better than one without ethics.
You must promote high ethical standards and create visibility to
discourage secrets and questionable deals.
Ethical standards keep your sales force honest.
This prevents destructive cynicism.
Good values are the only basis for all long-term success.

Sometimes you are forced to defend a salesperson's mistake.
As a sales manager, you must be honest about the situation.
You must let strong principles guide you.

Strong principles give your sales force an advantage.
Principles creates unity, which is the source of strength.

Stop the march when the rain swells the river into rapids. 3
You may want to ford the river.
Wait until it subsides.

4All regions can have seasonal mountain streams that can
cut you off.
There are seasonal lakes.
There are seasonal blockages.
There are seasonal jungles.
There are seasonal floods.
There are seasonal fissures.
Get away from all these quickly.
Do not get close to them.
Keep them at a distance.
Maneuver the enemy close to them.
Position yourself facing these dangers.
Push the enemy back into them.

16Danger can hide on your army's flank.
There are reservoirs and lakes.
There are reeds and thickets.
There are mountain woods.
Their dense vegetation provides a hiding place.
You must cautiously search through them.
They can always hide an ambush.

3 Stop your salespeople from tackling obstacles unnecessarily.
A salesperson tends to rush directly ahead.
Your job is sometimes to counsel patience.

Everyone on your sales force will eventually encounter a temporary obstruction.
There are naysayers.
There are gatekeepers.
There are time wasters.
There are deal killers.
There are lawyers.
Teach your salespeople to get around these obstacles quickly.
A sales force cannot invest sales time on professional obstacles.
Discourage sales relationships with such people.
Encourage salespeople to leave an opening for competitors.
Make your salespeople identify these sales barriers.
Make sure your competitors get bogged down with them.

Sales managers must identify hidden threats to sales.
You must be aware of overstocks.
You must be aware of problem agreements.
You must be aware of unhappy customers.
These difficulties can provide a base for competitive attack.
You must identify and address these problems.
Deny competitors ammunition against your sales organization.

Sometimes, the enemy is close by but remains calm. 4
Expect to find him in a natural stronghold.
Other times he remains at a distance but provokes battle.
He wants you to attack him.

5He sometimes shifts the position of his camp.
He is looking for an advantageous position.

7The trees in the forest move.
Expect that the enemy is coming.
The tall grasses obstruct your view.
Be suspicious.

11The birds take flight.
Expect that the enemy is hiding.
Animals startle.
Expect an ambush.

15Notice the dust.
It sometimes rises high in a straight line.
Vehicles are coming.
The dust appears low in a wide band.
Foot soldiers are coming.
The dust seems scattered in different areas.
The enemy is collecting firewood.
Any dust is light and settling down.
The enemy is setting up camp.

4 Sales managers must continually monitor competitors.
Competitors are confident in their most secure accounts.
Prevent your salespeople from attacking these competitors.
Attacking entrenched competition discredits your salespeople.

Competitors may shift their market focus.
They are looking for new opportunities.

Notice similar changes across different sales territories.
This means that a competitive threat is building.
You may not see which competitor is behind this threat.
You must be suspicious.

Notice when buyers suddenly become shy.
Suspect that competitors are planning a surprise.
Your customers are worried.
A competitor is challenging you.

Probe salespeople for rumors about competitors.
Salespeople learn about competitors from top decision-makers.
This foreshadows an aggressive move.
Salespeople hear news of competitors from many low-level people.
This means competitors have many people working the market.
News of competitors is scattered in different territories.
This means competitors are cherry-picking accounts.
News of competitors becomes rarer and rarer.
This means that competitors are dormant.

Your enemy speaks humbly while building up forces. *5*
He is planning to advance.

³The enemy talks aggressively and pushes as if to advance.
He is planning to retreat.

⁵Small vehicles exit his camp first.
They move the army's flanks.
They are forming a battle line.

⁸Your enemy tries to sue for peace but without offering a
treaty.
He is plotting.

¹⁰Your enemy's men run to leave and yet form ranks.
You should expect action.

¹²Half his army advances and the other half retreats.
He is luring you.

¹⁴Your enemy plans to fight but his men just stand there.
They are starving.

¹⁶Those who draw water drink it first.
They are thirsty.

¹⁸Your enemy sees an advantage but does not advance.
His men are tired.

5 Your competitors sound pessimistic but are hiring salespeople.
Prepare your salespeople for a new competitive campaign.

Your competitors make silly claims and talk about new markets.
Prepare your salespeople for competitors consolidating business.

Your competitors make quick changes.
They reorganize their product offering.
Prepare your salespeople for a competitive sales push.

Your competitors claim that they want to ally themselves with your company without a clear plan.
Prepare your salespeople for a betrayal.

Your competitors send mixed messages about their intentions.
Prepare for them to shift directions.

Competitors withdraw in some areas and move forward in others.
They are trying to outmaneuver your sales force.

Your competitors start offering unreasonable discounts.
Teach salespeople to portray competitors as desperate.

Your competitors start asking for cash payments.
Teach salespeople to question those competitors' stability.

Your competitors have a clear opportunity but do not pursue it.
They have overstretched their resources.

[20]Birds gather.
Your enemy has abandoned his camp.

[22]Your enemy's soldiers call in the night.
They are afraid.

[24]Your enemy's army is raucous.
The men do not take their commander seriously.

[26]Your enemy's banners and flags shift.
Order is breaking down.

[28]Your enemy's officers are irritable.
They are exhausted.

[30]Your enemy's men kill their horses for meat.
They are out of provisions.

[32]They don't put their pots away or return to their tents.
They are desperate.

[34]Enemy troops appear sincere and agreeable.
But their men are slow to speak to each other.
They are no longer united.

[37]Your enemy offers too many incentives to his men.
He is in trouble.

[39]Your enemy gives out too many punishments.
His men are weary.

Customers flock to your product.
Your competitors have abandoned some section of the market.

Competitors' salespeople ask your salespeople for information.
Teach your salespeople to recognize this as a sign of fright.

Competitors' salespeople are making outrageous claims.
Teach your salespeople that they are not to be taken seriously.

Competitors are undergoing reorganization.
Teach your salespeople to take advantage of their disorder.

Your competitors' best salespeople are upset.
Teach your salespeople to take advantage of their frustration.

Your competitors cut back on their travel budgets.
Teach your salespeople to take advantage of their restrictions.

Your competitors stop taking care of their customers.
Teach your salespeople to take advantage of their missteps.

Your competitors seem professional and dedicated.
They have poor communication within their company.
Teach your salespeople to take advantage of internal divisions.

Your competitors offer too many incentives to buy.
Teach your salespeople that this means that they are in trouble.

Your competitors start giving customers time limits.
Teach your salespeople that customers don't like pressure.

⁴¹Your enemy first acts violently and then is afraid of your
larger force.
His best troops have not arrived.

⁴³Your enemy comes in a conciliatory manner.
He needs to rest and recuperate.

⁴⁵Your enemy is angry and appears to welcome battle.
This goes on for a long time, but he doesn't attack.
He also doesn't leave the field.
You must watch him carefully.

If you are too weak to fight, you must find more men. 6
In this situation, you must not act aggressively.
You must unite your forces.
Prepare for the enemy.
Recruit men and stay where you are.

⁶You must be cautious about making plans and adjust to the
enemy.
You must gather more men.

Your competitors first attack your products and then try to ally with your company.

Teach your salespeople that this means that they are delaying.

Competitors try to collaborate with your salespeople.

Teach your salespeople that this means that they need help.

Competitors sound aggressive and seem to want your business.

They are always in your market but don't go after your customers.

They never lose their customers either.

Teach your salespeople to keep an eye out for these competitors.

6 A sales manager must grow the sales force when it is too small. When your sales force is overstretched, you cannot increase sales.

You must rally your current salespeople.

You prepare against competitors.

You must hire new salespeople but wait to expand.

You must be careful about opening new territories while looking for expansion opportunities.

Your new salespeople must first be trained.

With new, undedicated soldiers, you can depend on 7
them if you discipline them.
They will tend to disobey your orders.
If they do not obey your orders, they will be useless.

⁴You can depend on seasoned, dedicated soldiers.
But you must avoid disciplining them without reason.
Otherwise, you cannot use them.

⁷You must control your soldiers with esprit de corps.
You must bring them together by winning victories.
You must get them to believe in you.

¹⁰Make it easy for people to know what to do by training
your people.
Your people will then obey you.
If you do not make it easy for people to know what to do,
you won't train your people.
Then they will not obey.

¹⁴Make your commands easy to follow.
You must understand the way a crowd thinks.

✦ ✦ ✦

7 You must monitor new, unproven salespeople closely and correct them instantly to train them properly.
Otherwise, they will violate established guidelines.
If they do not understand your sales process, they cannot succeed.

You must treat your proven, established salespeople differently.
You must avoid routinely questioning their decisions.
If you do, you will lose them.

You must instill a spirit of teamwork in your salespeople.
You unite salespeople by winning sales through teamwork.
The basis of their faith is their success.

Make it easy for your salespeople to know what you expect when managing them.
They will then follow you.
If you make it too complicated for your salespeople to know what you want, you cannot manage them.
They will stop listening to you.

You must make it easy for salespeople to sell.
You must understand how competitors and customers think.

♦ ♦ ♦

Chapter 10

地 形

Field Position: Guiding Salespeople

Managing salespeople is like herding cats. It requires a strategic overview of your situation. As a sales manager, you must understand your salespeople's situation well enough to direct them. You must also be able to make judgments about why they succeed and why they fail. This chapter examines a wide variety of issues involved in providing the proper guidance to your sales force.

Good management starts with knowing what types of sales your salespeople are involved in. You not only want to make sure that they are spending their time with the best possible prospects, but you also need to be sure that they are using the best possible approach. Not all sales situations are the same. Sun Tzu's description of the six types of field positions provides a template for recognizing basic types of sales opportunities. It also teaches you how to respond appropriately to those opportunities.

Sales requires a certain type of character. Six character flaws can undermine any sales effort. You need to quickly diagnose these problems before you bring any salespeople on board.

It takes many skills to manage salespeople. You must have complete control of your sales program. When things go right, you must give the credit to the salespeople. When things go wrong, you will get the blame. A sales manager must be both tough and caring. You must challenge your salespeople to get the most out of them. You must train salespeople in every aspect of the sale: presenting the product, overcoming objections, and handling customers.

Field Position

SUN TZU SAID:

Some field positions are unobstructed. 1
Some field positions are entangling.
Some field positions are supporting.
Some field positions are constricted.
Some field positions give you a barricade.
Some field positions are spread out.

7You can attack from some positions easily.
Other forces can meet you easily as well.
We call these unobstructed positions.
These positions are open.
On them, be the first to occupy a high, sunny area.
Put yourself where you can defend your supply routes.
Then you will have an advantage.

Guiding Salespeople

1 Know which sales situations are open.
Know which sales situations are limited.
Know which sales situations are predictable.
Know which sales situations depend on fast reactions.
Know which sales situations are susceptible to creating barriers.
Know which sales situations are a poor fit for your company.

In some situations, customers readily accept your salespeople.
These customers will accept your competitors just as easily.
These are open sales situations.
These sales can go in any direction.
Instruct your salesperson to quickly grab the spotlight.
Make sure that your company treats these customers well.
These situations are great opportunities.

¹⁴You can attack from some positions easily.
Disaster arises when you try to return to them.
These are entangling positions.
These field positions are one-sided.
Wait until your enemy is unprepared.
You can then attack from these positions and win.
Avoid a well-prepared enemy.
You will try to attack and lose.
Since you can't return, you will meet disaster.
These field positions offer no advantage.

²⁴You cannot leave some positions without losing an advantage.
If the enemy leaves this ground, he also loses an advantage.
We call these supporting field positions.
These positions strengthen you.
The enemy may try to entice you away.
Still, hold your position.
You must entice the enemy to leave.
You then strike him as he is leaving.
These field positions offer an advantage.

³³Some field positions are constricted.
Get to these positions first.
You must fill these areas and await the enemy.
Sometimes, the enemy will reach them first.
If he fills them, do not follow him.
However, if he fails to fill them, you can go after him.

In some situations, your salespeople can get into accounts.
The problems arise when they try to sell them more.
These are limited sales situations.
These sales situations favor only one party.
You must counsel patience to your salespeople.
If they can get a profitable one-time deal, they can take it.
You must have them avoid well-trained buyers.
These buyers can take advantage of salespeople.
Since you won't get another order, the sale can cost you money.
These sales situations offer no strategic advantage.

In some sales situations, your salespeople cannot be inconsistent
without damaging the relationship.
You know competitors hurt themselves by being undependable.
These sales situations require predictability.
These relationships improve your company's operations.
Your salespeople may want to change an agreement.
You must not let them.
You must force salespeople to leave well enough alone.
Get rid of salespeople who cannot maintain these relationships.
These sales relationships are extremely valuable.

Some sales situations depend solely on fast reactions.
Make your salespeople respond instantly.
Your salespeople should satisfy the buyer before competitors arrive.
Your salespeople might not find out about these sales immediately.
If competitors have responded, your salespeople are wasting their time.
If competitors have not responded, your salespeople can still win.

39Some field positions give you a barricade.
Get to these positions first.
You must occupy their southern, sunny heights in order to
await the enemy.
Sometimes the enemy occupies these areas first.
If so, entice him away.
Never go after him.

45Some field positions are too spread out.
Your force may seem equal to the enemy.
Still you will lose if you provoke a battle.
If you fight, you will not have any advantage.

49These are the six types of field positions.
Each battleground has its own rules.
As a commander, you must know where to go.
You must examine each position closely.

Some armies can be outmaneuvered. 2
Some armies are too lax.
Some armies fall down.
Some armies fall apart.
Some armies are disorganized.
Some armies must retreat.

7Know all six of these weaknesses.
They create weak timing and disastrous positions.
They all arise from the army's commander.

In some sales situations, salespeople can erect competitive barriers.
Your salespeople must get to these decision-makers first.
Show your salespeople how to create the specifications that make competition difficult.
Sometimes competitors get into these situations first.
If so, check whether specifications can be changed.
Do not let your salespeople meet a competitor's specifications.

Some sales situations are a poor fit for your company.
A salesperson may think your company can meet the customer's needs.
Your job is to know when you cannot make it work.
Even if the salesperson makes the sale, it won't be profitable.

These are the six general types of sales situations.
Each sales prospect is a little different.
As the sales manager, you must guide your salespeople.
You must get them to pursue the best opportunities.

2 There are salespeople who are not problem solvers.
There are salespeople who are too sloppy.
There are salespeople who are too lazy.
There are salespeople who are divisive.
There are salespeople who are undisciplined.
There are salespeople who never win.

You must identify these six weaknesses in your salespeople.
These character flaws create problems in even the best accounts.
They all arise from a salesperson's character.

¹⁰One general can command a force equal to the enemy.
Still his enemy outflanks him.
This means that his army can be outmaneuvered.

¹³Another can have strong soldiers but weak officers.
This means that his army is too lax.

¹⁵Another has strong officers but weak soldiers.
This means that his army will fall down.

¹⁷Another has subcommanders that are angry and defiant.
They attack the enemy and fight their own battles.
The commander cannot know the battlefield.
This means that his army will fall apart.

²¹Another general is weak and easygoing.
He fails to make his orders clear.
His officers and men lack direction.
This shows in his military formations.
This means that his army is disorganized.

²⁶Another general fails to predict the enemy.
He pits his small forces against larger ones.
His weak forces attack stronger ones.
He fails to pick his fights correctly.
This means that his army must retreat.

Your salespeople have the same opportunites as the competition.
Your competitors always construct a better agreement.
These salespeople are poor problem solvers.

Some salespeople have good work habits but poor organization.
These salespeople are too sloppy.

Some salespeople have good organization but poor work habits.
These salespeople are too lazy.

Some salespeople are emotional and self-centered.
They attack competitors and think they know everything.
As a sales manager, you never know what they are doing.
These salespeople are divisive.

Some salespeople are vacillating and indulgent.
These salespeople cannot communicate company policy.
Their customers don't know what they can do.
This shows in the way the salespeople's sales are organized.
These salespeople are undisciplined.

Some salespeople cannot understand their competitive position.
These salespeople always see competitors as superior.
They mimic competitors instead of dominating them.
They fail to focus on the right customers.
These salespeople never win.

³¹You must know all about these six weaknesses.
You must understand the philosophies that lead to defeat.
When a general arrives, you can know what he will do.
You must study each general carefully.

You must control your field position. 3
It will always strengthen your army.

³You must predict the enemy to overpower him and win.
You must analyze the obstacles, dangers, and distances.
This is the best way to command.

⁶Understand your field position before you go to battle.
Then you will win.
You can fail to understand your field position and still fight.
Then you will lose.

¹⁰You must provoke battle when you will certainly win.
It doesn't matter what you are ordered.
The government may order you not to fight.
Despite that, you must always fight when you will win.

¹⁴Sometimes provoking a battle will lead to a loss.
The government may order you to fight.
Despite that, you must avoid battle when you will lose.

A sales manager must quickly recognize these six faults.
You must know the attitudes that lead to them.
When salespeople are hired, you must know that they can perform.
You must examine each prospective salesperson carefully.

3 A sales manager must determine the right sales process.
This always makes your sales force stronger.

You must teach salespeople how to counter competitors' claims.
You must teach how to address objections, problems, and pricing.
This is the best way to manage a sales department.

Salespeople must understand the sales process before they sell.
Then they will win consistently.
Salespeople may not know the sales process and try to sell.
Then they will often fail.

You must target accounts that your salespeople can easily satisfy.
It doesn't matter what other parts of your organization want.
Marketing or management may not agree.
If your salespeople win sales and make quota, all will be forgiven.

You must not waste time on sales that your salespeople cannot win.
Marketing or management may want you to change your focus.
If your department does not make quota, you will have no defense.

[17]You must advance without desiring praise.
You must retreat without fearing shame.
The only correct move is to preserve your troops.
This is how you serve your country.
This is how you reward your nation.

Think of your soldiers as little children. 4
You can make them follow you into a deep river.
Treat them as your beloved children.
You can lead them all to their deaths.

[5]Some leaders are generous but cannot use their men.
They love their men but cannot command them.
Their men are unruly and disorganized.
These leaders create spoiled children.
Their soldiers are useless.

You may know what your soldiers will do in an attack. 5
You may not know if the enemy is vulnerable to attack.
You will then win only half the time.
You may know that the enemy is vulnerable to attack.
You may not know if your men have the capability of attack-
ing him.
You will still win only half the time.
You may know that the enemy is vulnerable to attack.
You may know that your men are ready to attack.
You may not, however, know how to position yourself in the
field for battle.
You will still win only half the time.

When your programs work, you should praise your salespeople.
When your programs fail, you should take the blame.
Good sales management requires keeping salespeople upbeat.
This is how you make your company successful.
This is how you make money for everyone.

4 Treat your salespeople like your children.
Your leadership should challenge them to do the impossible.
You must put their own success first.
You must enable them to do their very best.

Softhearted sales managers cannot guide their salespeople.
Caring about people is not the same as managing them.
Undisciplined and confused salespeople cannot be successful.
Easygoing sales managers ruin their salespeople.
They create only failures.

5 You can train your salespeople how to present the product.
You must also teach your salespeople to deal with objections.
If you don't, you are wasting sales resources.
You can teach your salespeople to deal with objections.
You must also know that your salespeople are capable of befriend-
ing customers.
If you don't, you are wasting sales resources.
You can teach your salespeople to deal with objections.
You can train your salespeople to make friends with customers.
You must also understand the sales situations your people are deal-
ing with.
If you don't, you are wasting sales resources.

[11]You must know how to make war.
You can then act without confusion.
You can attempt anything.

[14]We say:
Know the enemy and know yourself.
Your victory will be painless.
Know the weather and the field.
Your victory will be complete.

† † †

You must know how to guide salespeople.
You must be confident in your decisions.
You can meet any sales goal.

Sales managers need information.
You need to know your competitors and your salespeople.
This makes selling easy.
You need to know the trends and the right market positions.
Then you can see your real opportunities.

✦ ✦ ✦

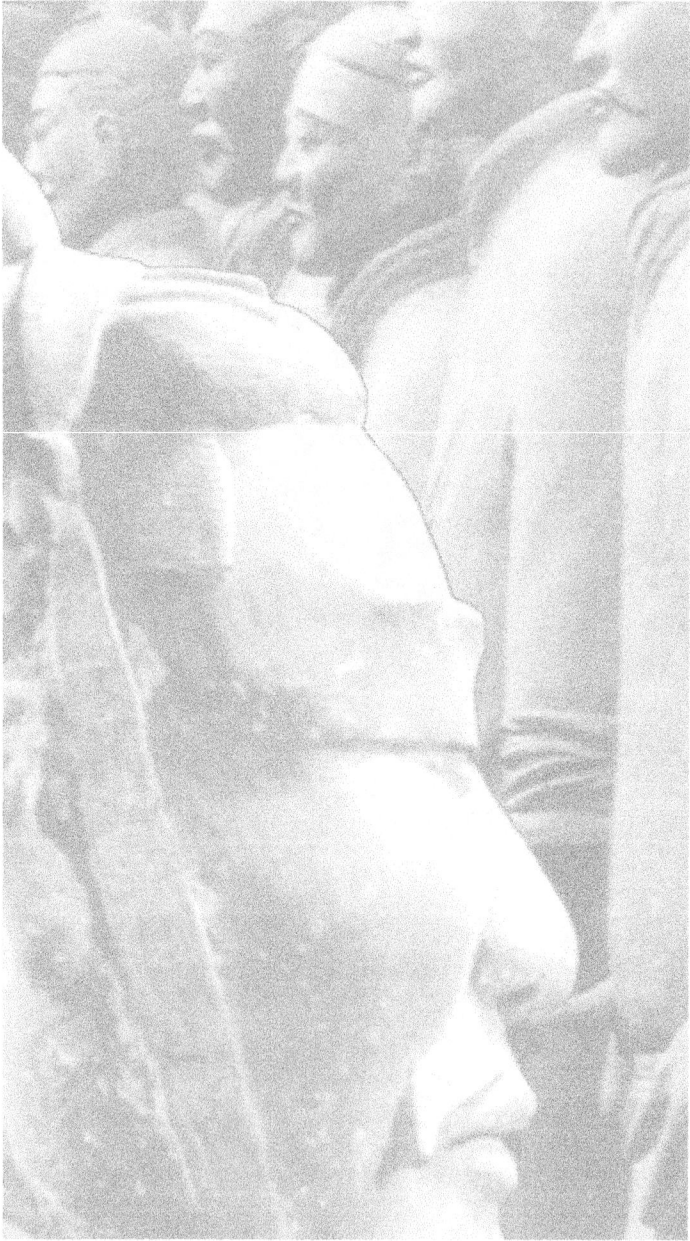

Chapter 11

九地

Types of Terrain: Management Challenges

As time goes on, your sales organization will face a wide variety of strategic challenges. Your salespeople will encounter problems in their territories. New competitors will come into your market. You will have to change your compensation systems. As a sales manager, you must not only immediately recognize these problems but know exactly how to respond to them.

First, you must recognize the different types of situations your salespeople face in their territories. Each territory is a little different and you must know how to best manage its specific conditions. Of course, since it is your responsibility to assign sales territories, you must also know how the boundaries are best drawn.

Successful sales management requires a healthy dose of psychology. You want your salespeople to be creative, independent individuals, but you also want them to work well in a group. To do this, the rules of strategy teach that you should use your competitors to unite and focus your sales force.

Your sales force must recognize your mastery of strategy. Salespeople don't have to love you or even like you, but they have to respect your vision. Your salespeople have to understand that it is your job to continually grow the business. You must use your sales force to gather competitive information and shape that information into competitive strategy. When you are threatened by new competitors, you must know how to rally the troops against them.

Types of Terrain

SUN TZU SAID:

Use the art of war. 1
Know when the terrain will scatter you.
Know when the terrain will be easy.
Know when the terrain will be disputed.
Know when the terrain is open.
Know when the terrain is intersecting.
Know when the terrain is dangerous.
Know when the terrain is bad.
Know when the terrain is confined.
Know when the terrain is deadly.

11Warring parties must sometimes fight inside their own territory.
This is scattering terrain.

13When you enter hostile territory, your penetration is shallow.
This is easy terrain.

15Some terrain gives you an advantageous position.
But it gives others an advantageous position as well.
This will be disputed terrain.

Management Challenges

1 Sales managers must react to sales challenges:
Recognize when your salespeople are in divisive situations.
Recognize when your salespeople are in easy situations.
Recognize when your salespeople are in competitive situations.
Recognize when your salespeople are in untapped situations.
Recognize when your salespeople are in shared situations.
Recognize when your salespeople are in risky situations.
Recognize when your salespeople are in difficult situations.
Recognize when your salespeople are in restricted situations.
Recognize when your salespeople are in desperate situations.

Salespeople are sometimes confused over whether or not a sales account is in their territory.
This is a divisive situation.

A new sales territory has had virtually no coverage.
This is an easy situation.

A sales territory has many good, established accounts in it.
These customer accounts also buy from your competition as well.
This is a competitive situation.

[18]You can use some terrain to advance easily.
Others can advance along with you.
This is open terrain.

[21]Everyone shares access to a given area.
The first one to arrive there can gather a larger group than
anyone else.
This is intersecting terrain.

[24]You can penetrate deeply into hostile territory.
Then many hostile cities are behind you.
This is dangerous terrain.

[27]There are mountain forests.
There are dangerous obstructions.
There are reservoirs.
Everyone confronts these obstacles on a campaign.
They make bad terrain.

[32]In some areas, the entry passage is narrow.
You are closed in as you try to get out of them.
In this type of area, a few people can effectively attack your
much larger force.
This is confined terrain.

[36]You can sometimes survive only if you fight quickly.
You will die if you delay.
This is deadly terrain.

Some sales territories have tremendous potential.
The competition, however, can still come in at any time.
This is an untapped situation.

Some customer accounts fall into several different territories.
The first salesperson to develop the accounts must bring in other salespeople to help.
This is a shared situation.

Successful salespeople can overextend themselves.
They will leave customers dissatisfied.
This is a risky situation.

There are problem customers.
There are accounts that withhold payment.
There are customers that want special treatment.
All salespeople deal with these people in their territory.
These are difficult situations.

There are some accounts that are difficult to develop.
To makes sales in these accounts you must meet their standards.
In these accounts, there are a number of people who can veto the purchase.
These are restricted situations.

Sometimes a salesperson needs immediate help with an account.
If the sales manager doesn't get involved, the customer will be lost.
These are desperate situations.

³⁹To be successful, you must control scattering terrain by
avoiding battle.
Control easy terrain by not stopping.
Control disputed terrain by not attacking.
Control open terrain by staying with the enemy's forces.
Control intersecting terrain by uniting with your allies.
Control dangerous terrain by plundering.
Control bad terrain by keeping on the move.
Control confined terrain by using surprise.
Control deadly terrain by fighting.

Go to an area that is known to be good for waging war. 2
Use it to cut off the enemy's contact between his front and
back lines.
Prevent his small parties from relying on his larger force.
Stop his strong divisions from rescuing his weak ones.
Prevent his officers from getting their men together.
Chase his soldiers apart to stop them from amassing.
Harass them to prevent their ranks from forming.

⁸When joining battle gives you an advantage, you must do it.
When it isn't to your benefit, you must avoid it.

¹⁰A daring soldier may ask:
"A large, organized enemy army and its general are coming.
What do I do to prepare for them?"

To be successful in divisive situations, prevent your salespeople
from fighting.
In easy territories, keep your salespeople working.
In competitive territories, salespeople must not attack competitors.
In untapped territories, make salespeople keep up with competitors.
In shared situations, teach salespeople how to work together.
In risky situations, focus salespeople on their best customers.
In difficult situations, make sure salespeople keep moving.
In restricting situations, force your salespeople to be inventive.
In desperate situations, involve yourself in the sale.

2 Sales managers must know how to divide sales territories.
By dividing your territories differently than competitors, you can
outmaneuver them.
You can put sales resources in areas competitors have neglected.
You can put your best salespeople against your competitors' weakest.
You can get more expertise into specialized marketplaces.
Discourage competitors from increasing their resources.
Discourage competitors from getting organized in your accounts.

You must create new territories when this generates new sales.
Do not create new territories when it doesn't generate new sales.

A good salesperson asks:
"A large, well-organized competitor is targeting my customers.
What should I do?"

¹³Tell him:
"First seize an area that the enemy must have.
Then he will pay attention to you.
Mastering speed is the essence of war.
Take advantage of a large enemy's inability to keep up.
Use a philosophy of avoiding difficult situations.
Attack the area where he doesn't expect you."

You must use the philosophy of an invader. 3
Invade deeply and then concentrate your forces.
This controls your men without oppressing them.

⁴Get your supplies from the riches of the territory.
It is sufficient to supply your whole army.

⁶Take care of your men and do not overtax them.
Your esprit de corps increases your momentum.
Keep your army moving and plan for surprises.
Make it difficult for the enemy to count your forces.
Position your men where there is no place to run.
They will then face death without fleeing.
They will find a way to survive.
Your officers and men will fight to their utmost.

¹⁴Military officers who are committed lose their fear.
When they have nowhere to run, they must stand firm.
Deep in enemy territory, they are captives.
Since they cannot escape, they will fight.

You must give your salespeople guidance.

They must go after one of that competitor's key customers.

Competitors respect salespeople who are dangerous.

Your salespeople must move more quickly than your competitors can.

Tell them to take advantage of larger competitors' slow reactions.

Prevent salespeople from getting into resource battles with competitors.

Instead, guide them into accounts that competitors take for granted.

3 Sales managers encourage their salespeople to be aggressive.

Insist that your salespeople develop special areas of expertise.

This narrows their focus without limiting their territory.

Base your salespeople's compensation on creating new sales.

The more you increase sales, the more sales support you can afford.

Make salespeople successful without overworking them.

Their individual success increases the whole company's momentum.

Keep everyone working hard and preparing for the worst.

You can make your organization seem much larger than it is.

Put salespeople in the position of having to close sales.

You cannot let them give up easily.

They will find a way to succeed.

Salespeople don't know what is possible until they try.

Salespeople who are talking with customers lose their fear.

When they are involved in sales, they forget their reluctance.

Keep them out in the field, calling on customers.

Since they have no choice, they will close sales.

[18]Commit your men completely.
Without being posted, they will be on guard.
Without being asked, they will get what is needed.
Without being forced, they will be dedicated.
Without being given orders, they can be trusted.

[23]Stop them from guessing by removing all their doubts.
Stop them from dying by giving them no place to run.

[25]Your officers may not be rich.
Nevertheless, they still desire plunder.
They may die young.
Nevertheless, they still want to live forever.

[29]You must order the time of attack.
Officers and men may sit and weep until their lapels are wet.
When they stand up, tears may stream down their cheeks.
Put them in a position where they cannot run.
They will show the greatest courage under fire.

Make good use of war. 4
This demands instant reflexes.
You must develop these instant reflexes.
Act like an ordinary mountain snake.
If people strike your head then stop them with your tail.
If they strike your tail then stop them with your head.
If they strike your middle then use both your head and tail.

Give your salespeople complete ownership of their territories.
They must feel responsible without you overseeing them.
They must solve problems without your intervention.
They must feel responsible to customers and not just to you.
You must be confident in their decision-making skills.

Make sales quotas, sales policies, and sales procedures clear.
Salespeople are more successful when you eliminate excuses.

Your salespeople may not be rich.
You must show them how they can all make serious money.
Your salespeople may fail.
You must show them how they can all be successful.

You alone must set sales quotas and sales deadlines.
Your salespeople will always complain and criticize.
Even when they accept your goals, they will still complain.
Keep salespeople busy closing sales.
This is when salespeople put their concerns behind them.

4 Sales managers use the competition.
You must train salespeople to respond instantly.
They must all immediately know the answer to every objection.
Sales flexibility comes from good reflexes.
You must prepare your salespeople for criticism of the company.
You must prepare them for challenges to your product.
You must prepare them for attacks on your services.

[8]A daring soldier asks:
"Can any army imitate these instant reflexes?"
We answer:
"It can."

[12]To command and get the most out of proud people, you
must study adversity.
People work together when they are in the same boat during
a storm.
In this situation, one rescues the other just as the right
hand helps the left.

[15]Use adversity correctly.
Tether your horses and bury your wagon's wheels.
Still, you can't depend on this alone.
An organized force is braver than lone individuals.
This is the art of organization.
Put the tough and weak together.
You must also use the terrain.

[22]Make good use of war.
Unite your men as one.
Never let them give up.

The commander must be a military professional. 5
This requires confidence and detachment.
You must maintain dignity and order.
You must control what your men see and hear.
They must follow you without knowing your plans.

Your salespeople may question prepared responses.

Can you prepare salespeople to answer every possible objection?

There is only one answer.

You must!

Sales managers get salespeople to help one another by rewarding the success of the group.

People work together best when a danger to one is dangerous to them all.

When you tie their success together, one salespereson will help another without thinking about it.

You must use competitive pressure.

You can lock salespeople into their territories.

This is not enough to make them successful.

Salespeople are better working together than working alone.

This requires cleverness in compensation.

Create teams that combine a range of skills.

You must know what your customers need.

Make good use of competition.

Make the success of the group important.

Use peer pressure.

5 Sales managers must be seen as experts in strategy.

Your salespeople must be confident in your impartiality.

Your salespeople must respect your authority.

You alone control your salespeople's perceptions of you.

Salespeople must trust that you know more than they do.

[6]You can reinvent your men's roles.

You can change your plans.

You can use your men without their understanding.

[9]You must shift your campgrounds.

You must take detours from the ordinary routes.

You must use your men without giving them your strategy.

[12]A commander provides what is needed now.

This is like climbing high and being willing to kick away your ladder.

You must be able to lead your men deeply into different surrounding territory.

And yet, you can discover the opportunity to win.

[16]You must drive men like a flock of sheep.

You must drive them to march.

You must drive them to attack.

You must never let them know where you are headed.

You must unite them into a great army.

You must then drive them against all opposition.

This is the job of a true commander.

[23]You must adapt to the different terrain.

You must adapt to find an advantage.

You must manage your people's affections.

You must study all these skills.

You can change your salespeople's assignments.

You can change your sales policies.

You must not seek permission from your salespeople.

You must occasionally change sales programs.

You must get salespeople out of their ruts.

Your salespeople can accept change without needing explanations.

Provide salespeople what they need when they need it.

You must continually ratchet up their objectives while removing any excuses for failure.

Salespeople must trust you enough to explore the possibilities of new markets.

This is the only way you discover expansion opportunities.

Guide your salespeople to keep them going in the right direction.

It is your responsibility to keep the sales department active.

You must keep salespeople looking for new business.

You must not set a final goal that makes it acceptable to quit prospecting.

You need salespeople to work together.

You must teach them to overcome all possible objections.

This is the job of a true sales manager.

You must adapt to every sales situation.

You must adjust your methods to win more sales.

You must control your customers' opinions.

You must master all these skills.

Always use the philosophy of invasion. 6
Deep invasions concentrate your forces.
Shallow invasions scatter your forces.
When you leave your country and cross the border, you must
take control.
This is always critical ground.
You can sometimes move in any direction.
This is always intersecting ground.
You can penetrate deeply into a territory.
This is always dangerous ground.
You penetrate only a little way.
This is always easy ground.
Your retreat is closed and the path ahead tight.
This is always confined ground.
There is sometimes no place to run.
This is always deadly ground.

[16]To use scattering terrain correctly, you must inspire your
men's devotion.
On easy terrain, you must keep in close communication.
On disputed terrain, you try to hamper the enemy's progress.
On open terrain, you must carefully defend your chosen position.
On intersecting terrain, you must solidify your alliances.
On dangerous terrain, you must ensure your food supplies.
On bad terrain, you must keep advancing along the road.
On confined terrain, you must stop information leaks from
your headquarters.
On deadly terrain, you must show what you can do by kill-
ing the enemy.

6 Sales managers must expand their market reach.
Meeting new challenges stimulates salespeople.
Dabbling in too many areas dissipates your sales force.
You must make your salespeople commit themselves to identifying a new segment.
This is an important decision.
You can sometimes go in several different directions.
Look for allies who share your goals.
You can invest heavily in winning a certain type of customer.
This is dangerous because it is expensive.
You are better off testing a niche market first.
This is much less risky.
Sometimes you don't have much of a choice of direction.
You are limited in your options.
Sometimes you have no place to go.
This is a desperate situation.

When you encounter a divisive situation, you must get salespeople to depend on one another.
In easy territories, make your salespeople report on their activities.
In competitive territories, focus salespeople on creating barriers.
In untapped territories, get salespeople to open key accounts.
In shared territories, make your salespeople work together.
In risky territories, give salespeople plenty of resources.
In difficult territories, have your salespeople find new markets.
In restricted territories, don't let your salespeople go around established procedures.
In desperate situations, your salespeople have no choice but to beat the competition.

²⁵Make your men feel like an army.
Surround them and they will defend themselves.
If they cannot avoid it, they will fight.
If they are under pressure, they will obey.

Do the right thing when you don't know your 7
different enemies' plans.
Don't attempt to meet them.

³You don't know the position of mountain forests, dangerous
obstructions, and reservoirs?
Then you cannot march the army.
You don't have local guides?
You won't get any of the benefits of the terrain.

⁷There are many factors in war.
You may lack knowledge of any one of them.
If so, it is wrong to take a nation into war.

¹⁰You must be able to control your government's war.
If you divide a big nation, it will be unable to put together a
large force.
Increase your enemy's fear of your ability.
Prevent his forces from getting together and organizing.

Make your salespeople proud of themselves.
You must not let salespeople use excuses to defend themselves.
You must make it easy for them to talk to customers.
You must challenge them if you expect to deserve your authority.

7 Sales managers must train their salespeople to know the competition's products and policies.
Salespeople cannot succeed without competitive knowledge.

Your salespeople must know about all possible objections, sales restrictions, and potential delivery problems.
If they don't, they cannot move forward.
Salespeople must have specialized knowledge.
If they don't, they cannot work with specific types of accounts.

Much information is required to sell a product.
Your job is to bring it all together.
It is wrong to manage a sales department without organizing it.

You alone must manage your company's sales policies.
If too many people determine sales policies, your sales force will be divided.
You must make your sales force a competitive threat.
Your work is to divide and disorganize your competitors.

¹⁴Do the right thing and do not arrange outside alliances
before their time.
You will not have to assert your authority prematurely.
Trust only yourself and your self-interest.
This increases the enemy's fear of you.
You can make one of his allies withdraw.
His whole nation can fall.

²⁰Distribute rewards without worrying about having a system.
Halt without the government's command.
Attack with the whole strength of your army.
Use your army as if it were a single man.

²⁴Attack with skill.
Do not discuss it.
Attack when you have an advantage.
Do not talk about the dangers.
When you can launch your army into deadly ground, even if
it stumbles, it can still survive.
You can be weakened in a deadly battle and yet be stronger
afterward.

³⁰Even a large force can fall into misfortune.
If you fall behind, however, you can still turn defeat into victory.
You must use the skills of war.
To survive, you must adapt yourself to your enemy's purpose.
You must stay with him no matter where he goes.
It may take a thousand miles to kill the general.
If you correctly understand him, you can find the skill to do it.

As sales manager, you must be cautious about the timing of sales alliances.

Wait until the right time to make commitments.

These arrangements must serve your company's best interests.

Your competitors must fear what you can do.

You might tempt away your competitors' allies.

This can completely undermine their plans.

Control compensation without worrying about pleasing everyone.

Stop programs that don't work, without debating it.

You must accentuate the strengths of your organization.

Everyone must share the same goals.

Make decisions like a leader.

Don't debate your judgments.

Expand when you identify an opportunity.

Do not emphasize what can go wrong.

You can find yourself in a risky situation and even make mistakes, but you can still succeed.

You can lose salespeople in a difficult period and become a better sales force for it.

Even the best sales forces get into trouble.

Think of each setback as a springboard for future success.

This requires a long-term strategy.

In bad times, you must recognize what competitors are doing right.

You must keep your company competitive no matter what.

You may have to transform your company to beat the competition.

If you know your competitors, you will eventually succeed.

Manage your government correctly at the start of a war. 8
Close your borders and tear up passports.
Block the passage of envoys.
Encourage politicians at headquarters to stay out of it.
You must use any means to put an end to politics.
Your enemy's people will leave you an opening.
You must instantly invade through it.

[8]Immediately seize a place that they love.
Do it quickly.
Trample any border to pursue the enemy.
Use your judgment about when to fight.

[12]Doing the right thing at the start of war is like
approaching a woman.
Your enemy's men must open the door.
After that, you should act like a streaking rabbit.
The enemy will be unable to catch you.

♦ ♦ ♦

8 Sales managers can reorganize to meet new competitors.
Keep proprietary information away from the competition.
Stop your salespeople from sharing.
Encourage everyone in your company to keep company secrets.
Every part of your company must support you.
Have your salespeople identify the weaknesses in new competitors.
Your salespeople must immediately start targeting them.

Direct your salespeople to go after a competitor's key accounts.
Waste no time.
Invent new policies to harass new competitors.
As the sales manager, you must pick the right battles.

You must be diplomatic in meeting the challenges of a new competitor.
Wait until the competitor gives you an opening.
After that, you should react quickly and unpredictably.
Your new competitor will be unable to keep up.

✦ ✦ ✦

Chapter 12

火攻

Attacking with Fire: Competitive Vulnerability

"Fire attacks" are not normal strategic attacks that hamper competitors, but direct attacks to hurt competitors. These attacks can easily backfire, destroying your salespeople's credibility, but they are a big part of competitive reality. Sales managers must take the responsibility for deciding when and what type of competitive attacks their salespeople should make.

Strategy teaches that you cannot manufacture these types of attacks out of thin air. Your opponents must provide the right targets and opportunity. Sun Tzu lists five targets to think about. The environment must also provide the proper conditions for a competitive attack to work.

Whether you are using these attacks yourself or are the target of them, everything depends on the target's reactions to them. There are five different attack scenarios that you must train your salespeople to recognize. These attacks seldom work exactly as planned. Using them means being prepared for what is likely to happen. Like so much of what strategy teaches, success depends on properly understanding your business environment.

Because competitive attacks are so dangerous, a sales manager's good judgment is the key to using them correctly. When used correctly, competitive attacks are an invaluable tool for winning sales. However, you must make sure that these attacks are aimed at winning sales and are not just launched out of hostility.

Attacking with Fire

SUN TZU SAID:

There are five ways of attacking with fire. 1
The first is burning troops.
The second is burning supplies.
The third is burning supply transport.
The fourth is burning storehouses.
The fifth is burning camps.

7To make fire, you must have the resources.
To build a fire, you must prepare the raw materials.

9To attack with fire, you must be in the right season.
To start a fire, you must have the time.

11Choose the right season.
The weather must be dry.

13Choose the right time.
Pick a season when the grass is as high as the side of a cart.

15You can tell the proper days by the stars in the night sky.
You want days when the wind rises in the morning.

Competitive Vulnerability

1 Sales managers must identify targets vulnerable to attack.
People can be targets.
Products can be targets.
Services can be targets.
Financial resources can be targets.
Organizations can be targets.

The environment makes your competition vulnerable to attack.
You must show your salespeople how to use these resources correctly.

The business climate determines when opponents are vulnerable.
You must guide your salespeople to use their time correctly.

Your salespeople cannot spark distrust unless the time is right.
The business climate must be ready.

As the sales manager, you must decide when the time is right.
Wait until there is plenty of fuel for your salespeople's accusations.

A good sales manager has a sense for when attacks will work.
It is conditions in the environment that make them successful.

Everyone attacks with fire. 2
You must create five different situations with fire and be able
to adjust to them.

³You start a fire inside the enemy's camp.
Then attack the enemy's periphery.

⁵You launch a fire attack, but the enemy remains calm.
Wait and do not attack.

⁷The fire reaches its height.
Follow its path if you can.
If you can't follow it, stay where you are.

¹⁰Spreading fires on the outside of camp can kill.
You can't always get fire inside the enemy's camp.
Take your time in spreading it.

¹³Set the fire when the wind is at your back.
Don't attack into the wind.
Daytime winds last a long time.
Night winds fade quickly.

¹⁷Every army must know how to adjust to the five possible
attacks by fire.
Use many men to guard against them.

2 All sales managers use attacks and must defend against them. The better you know how to use the five attack scenarios, the better you can protect yourself.

You aim competitive attacks at the core of an opponent's business.
Your salespeople use the charges to win over peripheral customers.

The secret to defending against competitive attacks is to keep calm.
If your opponents don't react, your salespeople cannot press the issue.

Sales attacks must fire the imagination and take on a life of their own.
Your salespeople can then win orders using them.
If attacks don't spark interest, your salespeople should forget them.

Your salespeople can also spread small doubts and questions.
This works when big targets are unavailable.
This approach is a long-term strategy.

These attacks work only when the business climate supports them.
Attacks always backfire when the climate is against you.
The more visible the trends are, the longer they last.
The more subtle they are, the more quickly they fade.

All of your salespeople must be trained to recognize these different situations and to respond appropriately.
They must always be on guard for competitive vulnerabilities.

When you use fire to assist your attacks, you are clever. 3
Water can add force to an attack.
You can also use water to disrupt an enemy.
It does not, however, take his resources.

You win in battle by getting the opportunity to attack. 4
It is dangerous if you fail to study how to accomplish this
achievement.
As commander, you cannot waste your opportunities.

4We say:
A wise leader plans success.
A good general studies it.
If there is little to be gained, don't act.
If there is little to win, do not use your men.
If there is no danger, don't fight.

10As leader, you cannot let your anger interfere with the suc-
cess of your forces.
As commander, you cannot let yourself become enraged
before you go to battle.
Join the battle only when it is in your advantage to act.
If there is no advantage in joining a battle, stay put.

14Anger can change back into happiness.
Rage can change back into joy.
A nation once destroyed cannot be brought back to life.
Dead men do not return to the living.

3 Leveraging the environment against opponents takes intelligence.
Changes in the environment can help your cause.

Change alone can hamper an opponent's progress.

Change alone, however, cannot destroy an opponent's business.

4 Sales managers succeed by devising attacks that work.
If you do not know how to use competitive vulnerabilities, you will
have a short career.

You must make full use of the few opportunities you will get.

The truth is simple.

Your competitive attacks must be organized.

Attacks work only if they are based on superior knowledge.

These attacks make sense only if they win customers for you.

Stop your salespeople from using attacks that fail to win customers.

If the competitor's weakness isn't real, you cannot use it.

As a sales manager, you must prevent your competitive hostility
from hurting your sales.

Set an example for your salespeople by not letting emotions affect
your judgment.

Salespeople must attack competitors only when it wins customers.

If it won't affect a sale, they should say nothing about competitors.

Hostility can change into friendship.

Today's competitors can become tomorrow's allies.

You cannot undo the damage of an attack once it is done.

You can destroy your own credibility.

[18]This fact must make a wise leader cautious.
A good general is on guard.

[20]Your philosophy must be to keep the nation peaceful and
the army intact.

✦ ✦ ✦

Because attacks are dangerous, a sales manager is careful.
You must always be on guard against them.

You make your salespeople successful when your ethics discourage
worthless conflict.

♦ ♦ ♦

Chapter 13

用 間

Using Spies: The Power of Information

To be successful as a sales manager, you must make your salespeople respect your knowledge. Your salespeople must never think that you are out of touch. Your information is only as good as its sources. One of your key responsibilities as a sales manager is to teach your salespeople to continually collect information for the organization. Beyond that, you must have contacts that provide you with information that you can't get from your salespeople.

Strategically, information is an economic resource. You have only limited time, effort, and funds. If you want good information, you will to have to spend some of those limited resources on collecting and organizing information. It is easy to overlook this part of management, but classical strategy insists that information is more valuable than anything else that you can spend these resources on.

When it comes to developing conduits for information, Sun Tzu teaches that there are five types of resources you need in your network. Four of these contacts collect specific types of information. One of them is used to provide misleading information to your opponents.

To properly manage information, you must get details from a variety of sources. Each individual point of view adds something valuable, but all of them are necessary to develop a complete picture of your situation. You especially need to know how to get inside your competitors' information networks to learn what they are thinking and planning.

Using Spies

SUN TZU SAID:

All successful armies require thousands of men. 1
They invade and march thousands of miles.
Whole families are destroyed.
Other families must be heavily taxed.
Every day, a large amount of money must be spent.

[6]Internal and external events force people to move.
They are unable to work while on the road.
They are unable to find and hold a useful job.
This affects seventy percent of thousands of families.

[10]You can watch and guard for years.
Then a single battle can determine victory in a day.
Despite this, bureaucrats worship the value of their salary
money too dearly.
They remain ignorant of the enemy's condition.
The result is cruel.

[15]They are not leaders of men.
They are not servants of the state.
They are not masters of victory.

The Power of Information

1 Sales managers are directly responsible for their salespeople.
You alone must keep all your salespeople moving forward.
Every salesperson is at risk.
Your salespeople will pay a heavy price for your mistakes.
Your company is investing heavily every day in your success.

Selling affects everyone in your company directly or indirectly.
If salespeople don't make sales, other employess will lose their jobs.
Bad sales can force many to leave to find other jobs.
The livelihood of all these people depends on your decisions.

People can work for years protecting and maintaining a company.
A single sales breakthrough can transform that company overnight.
Despite this, too many sales managers are satisfied with their salary and their position.
These sales managers fail to clearly understand their opportunities.
They are failures.

These sales managers are not managing salespeople.
These sales managers are not serving their companies.
These sales managers are not even marginally successful.

[18]You need a creative leader and a worthy commander.
You must move your troops to the right places to beat others.
You must accomplish your attack and escape unharmed.
This requires foreknowledge.
You can obtain foreknowledge.
You can't get it from demons or spirits.
You can't see it from professional experience.
You can't check it with analysis.
You can only get it from other people.
You must always know the enemy's situation.

You must use five types of spies. 2
You need local spies.
You need inside spies.
You need double agents.
You need doomed spies.
You need surviving spies.

[7]You need all five types of spies.
No one must discover your methods.
You will then be able to put together a true picture.
This is the commander's most valuable resource.

[11]You need local spies.
Get them by hiring people from the countryside.

[13]You need inside spies.
Win them by subverting government officials.

You must become an inventive and brilliant sales manager.
You must inspire your salespeople to dominate your market.
You must win new markets without hurting profitability.
This requires the right information.
You can get this information.
You won't get it from sales forecasts.
You won't get it from your past experience as a sales manager.
You can't reason out what you need to learn.
You get information only by developing contacts with other people.
You must always learn where your opportunities lie.

2 There are only five types of information sources.
There are contacts in your marketplace.
There are contacts close to key customers.
There are contacts who provide information to your competitors.
There are contacts that you can sacrifice.
There are contacts that you must keep alive.

You must use all five types of information sources.
If you do, no one will ever challenge your knowledge.
You can discover where those breakthrough opportunities are.
This information is your most valuable resource.

You need to know your marketplace.
You must have good sources throughout your target market.

You need to know your key customers.
You must win over other people in those organizations.

¹⁵You need double agents.
Discover enemy agents and convert them.

¹⁷You need doomed spies.
Deceive professionals into being captured.
Let them know your orders.
They then take those orders to your enemy.

²¹You need surviving spies.
Someone must return with a report.

Your job is to build a complete army. 3
No relations are as intimate as the ones with spies.
No rewards are too generous for spies.
No work is as secret as that of spies.

⁵If you aren't clever and wise, you can't use spies.
If you aren't fair and just, you can't use spies.
If you can't see the small subtleties, you won't get the truth
from spies.

⁸Pay attention to small, trifling details!
Spies are helpful in every area.

¹⁰Spies are the first to hear information, so they must not
spread information.
Spies who give your location or talk to others must be killed
along with those to whom they have talked.

You need information on your competitors.
You must find a way to tap into their communication channels.

You need contacts you can sacrifice.
You want your worst salespeople to move to your competitors.
Give them information that makes them seem valuable.
You can get rid of them while misleading your competition.

You need contacts with whom you can build lasting relationships.
These are people who get critical information when you need it.

3 A sales manager must create a powerful sales force.
You want your best information sources to be your closest friends.
You must reward those who bring you the best information.
You must know how to keep information confidential.

Your information network will fall apart if you use it foolishly.
Your information network will fall apart if you abuse it.
Your information network will mislead you if you ignore the fine
points of detail.

Tidbits of information can be the most valuable.
Cultivate a variety of communication channels.

You want sources of information who know how to keep a confidence and who talk only to you.
You must cut off information sources who discuss your plans with
others.

You may want to attack an army's position. 4
You may want to attack a certain fortification.
You may want to kill people in a certain place.
You must first know the guarding general.
You must know his left and right flanks.
You must know his hierarchy.
You must know the way in.
You must know where different people are stationed.
You must demand this information from your spies.

10You want to know the enemy spies in order to convert
them into your men.
You find sources of information and bribe them.
You must bring them in with you.
You must obtain them as double agents and use them as
your emissaries.

14Do this correctly and carefully.
You can contact both local and inside spies and obtain their
support.
Do this correctly and carefully.
You create doomed spies by deceiving professionals.
You can use them to give false information.
Do this correctly and carefully.
You must have surviving spies capable of bringing you
information at the right time.

4 Sales managers must get specific information on key sales.
Your salespeople may want to win a certain account.
You may want to push competitors out of a certain market.
You must first know who the key decision-makers are.
You must understand who influences decisions.
You must know their organization.
You must know the best ways to reach them.
You must understand different people's responsibilities.
You must get this information from your salespeople.

You must instruct your salespeople to win over people who provide information to your opponents.
Give your salespeople rewards for developing these contacts.
Your salespeople must win support from friends of your opponents.
The more positive relationships your organization develops among an opponent's supporters, the stronger it is.

Teach your salespeople to be cautious.
You need reliable information in target markets and within key accounts.
Discourage salespeople from believing everything they hear.
You want competitors to hire away your worst salespeople.
You can use them to influence your competitors' thinking.
This is a delicate matter.
You must have lasting relationships with people who know what you need to know before you do.

²¹These are the five different types of intelligence work.
You must be certain to master them all.
You must be certain to create double agents.
You cannot afford to be too cost conscious in creating these double agents.

This technique created the success of ancient Shang. 5
This is how the Shang held its dynasty.

³You must always be careful of your success.
Learn from Lu Ya of Shang.

⁵Be a smart commander and a good general.
You do this by using your best and brightest people for spying.
This is how you achieve the greatest success.
This is how you meet the necessities of war.
The whole army's position and ability to move depends on these spies.

There are five different types of information conduits.
As a sales manager, you must master them all.
You must be certain to get inside your opponents' thinking.
You cannot invest too much time, effort, and money in understanding your opponents' priorities.

5 All great sales organizations need an information network.
This is how they succeed year after year.

You can ensure that your sales department is successful.
Learn from the success of others.

Your salespeople need you to be informed and knowledgeable.
The best salespeople are those who value and share information.
Sales breakthroughs come only from great information.
This is how you stay competitive in the marketplace.
Your whole company's market position and your sales department's success depend on information.

✦ ✦ ✦

✦ ✦ ✦

About the Translator and Author

Gary Gagliardi is recognized as America's leading expert on Sun Tzu's *The Art of War*. An award-winning author and businessperson, he is known for his ability to put sophisticated concepts into simple, easy-to-understand terms. He appears on hundreds of talk shows nationwide, providing strategic insight on the breaking news.

Gary began studying the Chinese classic more than thirty years ago, applying its principles first to his own career, then to building a successful business, and finally in training the world's largest organizations to be more competitive. He has spoken all over the world on a variety of topics concerning competition, from modern technology to ancient history. His books have been translated into many languages, including Japanese, Korean, Russian, and Spanish.

Gary began using Sun Tzu's competitive principles in a successful corporate career but soon started his own software company. In 1990, he wrote his first *Art of War* adaptation for his company's salespeople. By 1992, his company was on *Inc.* magazine's list of the 500 fastest-growing privately held companies in America. After he won the U.S. Chamber of Commerce **Blue Chip Quality Award** and became an Ernst and Young **Entrepreneur of the Year** finalist, Gary's customers—AT&T, GE, and Motorola, among others—began inviting him to speak at their conferences. Jardin's, the original Hong Kong trading company known as "The Noble House," became one of his partners and even gave him the honor of firing the noontime cannon in Hong Kong's harbor. After becoming a multimillionaire when he sold his software company in 1997, he continued teaching *The Art of War* around the world.

Gary has authored several breakthrough works on *The Art of War*. In 1999, he translated each Chinese character to demonstrate the system's equation-like symmetry. In 2003, his work *The Art of War Plus The Ancient Chinese Revealed* won the **Independent Publishers Book Award** as the year's best new multicultural nonfiction work. In 2004, his adaptation of Sun Tzu's principles to marketing, *The Art of War Plus The Art of Marketing*, was selected as one of the year's three best business books by the **Ben Franklin Book Awards committee**. In 2004, he released a new work that explains the many hidden aspects of Sun Tzu's text, *The Art of War Plus Its Amazing Secrets*, which was selected as a **Highlighted Title** by Independent Publishers.

Gary has also written a large number of other adaptations of *The Art of War*, applying Sun Tzu's methods to areas such as career building, management, small business, and even romance and parenting.

Art of War Gift Books

Mastering Strategy Series

^{Sun Tzu's} The Art of War Plus The Warrior's Apprentice
A first book on strategy for the novice.

^{Sun Tzu's} The Art of War Plus The Ancient Chinese Revealed
See the original! Each original Chinese character individually translated.

^{Sun Tzu's} The Art of War Plus Its Amazing Secrets
Learn the hidden secrets! The deeper meaning of Sun Tzu explained.

The Warrior Class: 306 Lessons in Strategy
The complete study guide! Simple lessons in strategy you can read any time.

Career and Business Series

^{Sun Tzu's} The Art of War Plus The Art of Career Building
For everyone! Use Sun Tzu's lessons to advance your career.

^{Sun Tzu's} The Art of War Plus The Art of Sales
For salespeople! Use Sun Tzu's lessons to win sales and keep customers.

^{Sun Tzu's} The Art of War Plus The Art of Management
For managers! Use Sun Tzu's lessons on managing teams more effectively.

^{Sun Tzu's} The Art of War Plus Strategy for Sales Managers
For sales managers! Use Sun Tzu's lessons to direct salespeople more effectively.

^{Sun Tzu's} The Art of War Plus The Art of Small Business
For business owners! Use Sun Tzu's lessons in building your own business.

^{Sun Tzu's} The Art of War Plus The Art of Marketing
For marketing professionals! Use Sun Tzu's lessons to win marketing warfare.

Life Strategies Series

^{Sun Tzu's} The Art of War Plus The Art of Love
For lifelong love! *Bing-fa* applied to finding, winning, and keeping love alive.

^{Sun Tzu's} The Art of War Plus The Art of Parenting Teens
For every parent! Strategy applied to protecting, guiding, and motivating teens.

Current Events Series

^{Sun Tzu's} The Art of War Plus Strategy against Terror
An examination of the War on Terror using Sun Tzu's timeless principles.

Audio and Video

Amazing Secrets of *The Art of War*: Audio
1 1/2 Hours 2-CD set

Amazing Secrets of *The Art of War*: Video
1 1/2 Hours VHS

**To Order On-line: Visit www.BooksOnStrategy.com
Fax Orders: 206-546-9756. Voice: 206-533-9357.**